T0138343

Galileo's Idol

Published with support of the Susan E. Abrams Fund

Galileo's Idol

Gianfrancesco Sagredo and the Politics of Knowledge

NICK WILDING

The University of Chicago Press
Chicago and London

Nick Wilding is assistant professor in the Department of History at Georgia State University.

The University of Chicago Press, Chicago 60637
The University of Chicago Press, Ltd., London
© 2014 by The University of Chicago
All rights reserved. Published 2014.
Printed in the United States of America

23 22 21 20 19 18 17 16 15 14   1 2 3 4 5

ISBN-13: 978-0-226-16697-1 (cloth)
ISBN-13: 978-0-226-16702-2 (e-book)
DOI: 10.7208/chicago/9780226167022.001.0001

Library of Congress Cataloging-in-Publication Data

Wilding, Nick, author.
    Galileo's idol : Gianfrancesco Sagredo and the politics of knowledge /
Nick Wilding.
        pages cm
    Includes bibliographical references and index.
    ISBN 978-0-226-16697-1 (cloth : alk. paper)—ISBN 978-0-226-16702-2 (e-book)
1. Sagredo, Gianfrancesco, 1571–1620. 2. Galilei, Galileo, 1564–1642. 3. Science—
Biography.   I. Title.
    Q143.S17W55 2014
    509.2—dc23
    [B]

                                                        2014012022

♾ This paper meets the requirements of ANSI/NISO Z39.48-1992 (Permanence of
Paper).

Contents

Introduction

Galileo's Idol is a historical case study of machinations, maneuvers, and masks in the making of early modern scientific knowledge. It pays special attention to the contributions of unacknowledged underground actors such as printers, publishers, forgers, diplomats, politicians and travelers. My book retells a familiar story, that of Galileo's transformation of the cosmos via the introduction of astronomical telescopic observation, from a new and, I hope, newly productive angle. Rather than taking Galileo as the central voice and protagonist in this narrative, it approaches him askance, from the position of his closest friend, student, and patron, the Venetian patrician Gianfrancesco Sagredo.[1] The raking light supplied by this approach not only casts long shadows, it also helps illuminate certain features not usually visible, and allows us to explore relationships between objects and practices rarely considered together. This tactic is not that of *Rosencrantz and Guildenstern are Dead* to *Hamlet*; Galileo still figures large in the narrative, but by coming at him sideways, we can both catch him unawares and observe him in his natural habitat, as it were. The historical contexts produced by and generating this line of inquiry are not limited to those of a single overprivileged relationship; my initial project of writing the history of an early modern scientific friendship, based on the rich documentation of Sagredo's letters to Galileo, has here been eclipsed by a series of complimentary moves based on unexpected discoveries and improvised methodologies. An account of the friendship and, indeed, an intellectual biography of Sagredo remain to be written.[2] Instead, this book attempts to capture something of the dynamics of early modern scientific practice in action. My aim is not to propose Sagredo's inclusion in some canon of scientific revolutionaries, but to contextualize Galileo's activities in a new way, by viewing him alongside his friend.

In the case of Galileo, there is no laboratory or observatory for us to enter, no space defined as scientific, even were such a term to have existed. When Galileo later imagined a proper space for talking about science, as the settings for his two great dialogues of 1632 and 1638, the best he could come up with was Sagredo's palace (situated not on the Grand Canal, where the family later moved, but up in Castello against the walls of the Arsenal shipyard). Other contemporaries did conceptualize and utilize such spaces, and they have been well studied.[3] Instead, in Galileo's case, we have to identify and follow the material, intellectual, and social tools that he and his assistants, friends, patrons, enemies, admirers, rivals, and students constructed and used. I am less concerned here with telling the story again of the physical construction of the telescope and the observational breakthroughs of the winter of 1609–10, and more interested in trying to reconstruct the processes by which other tools and instruments were made and deployed. I aim to restore something of the social distribution of labor and historical contingency in the making of natural philosophical knowledge.

To this end, I have used as wide a variety of sources as possible. Paintings, ornamental woodcuts, epistolary hoaxes, intercepted letters, murder case files: these are not the normal materials for historians of science. This microhistorical case study is intended to investigate an abnormality, or rather, it is intended to challenge the picture that has emerged as dominant and normative, of early modern science as pious, serious, and ecumenical—one might say, as institutionalized. So powerful and perceptive were the original social constructivist studies which provided this picture, especially around central figures of the early Royal Society, that we have forgotten, against their authors' wishes, that they were also intended as microhistorical analyses, inherently local and richly descriptive, rather than prescriptive checklists for a new generation.[4]

Galileo emerges from this book as arrogant, manipulative, and scheming. This much was well known. These adjectives should not be understood as psychological traits, but as cultural resources, borrowed, deployed, and valued across a wide range of activities. Such resources are not amassed capital, but are generated in performance. As such, they depend on an audience, or rather, they are generated relationally. Whereas a natural philosopher such as Boyle might plead for toleration and moderation to produce nondogmatic facts, or Hobbes might insist on the right and need for an authoritative voice to stabilize truth, the cases here presented, emanating from a different space, time, and culture, show instead a divisive approach to the art of persuasion. These authors perform for modulated audiences, in curious contracts with variegated interpretative communities. Hoaxes are one of their most highly

prized productions, texts that simultaneously appear absolutely sincere to
one group and outrageously funny to another—funny precisely because
someone else fails to see the humor. Galileo's *Il Saggiatore* is perhaps the
best example of science in a satiric mode; his *Dialogo* the most complex, as
it writes its victim into its plot. Natural philosophy takes place in fields, in
dialect, behind masks, at carnival. It is polemical, rude, satirical, and enjoyable; it insults, misrepresents, tricks, and ridicules. Even when its setting is a
Venetian palazzo, it nearly turns into a fistfight. Only later does it enter the
pious space of academies, where monastic and courtly codes of dullness are
imposed; and even there, we may have been too quick to miss the jokes.

Such works do not rely on a notion of credibility that is socially sanc-
tioned, a fixed resource from which the truth of an utterance draws its
strength. Rather, they are complex textual performances whose meaning is
generated through a series of negotiations and partial releases of information. The early modern "discreet reader" was ideal but also real. Discernment
in locating an author's motives, subterfuges, identity, or humor was the same
skill that allowed one to penetrate the secrets of nature. Wit, rather than sincerity, was the highest virtue of this epistemology. Trust was not guaranteed
by an author's name: it was perfectly possible for a pseudonym or an anonym
to be more truthful than an onym.

Similarly, strong studies in the relationship between print and science
have inadvertently created a normative procrustean bed into which quite different social relationships have been forced.[5] The cases offered here show
very clearly that the printer of Galileo's *Sidereus nuncius* (Venice, 1610) did
not even place his name on its title page, but hid behind a series of masks
from a very real threat. The supposed "pirate" second edition from Frankfurt, by contrast, was by a highly reputable publisher. Both editions were
probably "authorial," but they served very different audiences. Readers did
not discuss which edition they used, but did understand the work in very different ways, depending, in part, on their familiarity with what they deemed
similar works. Furthermore, the primacy of print, a point of faith since at
least Francis Bacon in its historical agency, no longer seems so secure. Letters
and manuscript tracts make up the bulk of most early modern natural philosophers' writing. For many texts they were the only, or privileged, or first,
form of publication.[6]

Patrons have generally been deployed as fixed rungs on a hierarchy up
which our aspiring scientist, artist, writer, engineer, or doctor might climb.[7]
Instead, it might make more sense to consider entire systems of patronage in
movement, with no fixed points, only shifting relations. This is more than a
restatement of the reciprocity of the gift; it is both an attempt to restore early

modern notions of contingency and a way of undoing the teleology of our notion of actors' strategies. Strategies have helpfully replaced intentions as a way of de-psychologizing actors and following behavior rather than mere words. In turn, though, the notion of strategy often reduces historical potentialities to their results, constructing linear and singular narratives with no space for failure, improvisation, or self-contradiction. The case studies offered here, by contrast, seek to restore multiple, contradictory tactics to actors. Galileo is here positioned alongside actors better known in the political arena, such as Paolo Sarpi. But he is also restored to a social context granting agency to amanuenses, printers, and publishers.

A nuanced notion of strategy allows us to reconsider some of the staple, even stodgy, components of Venetian intellectual history. Anti-Jesuitism can be studied and described with more sensitivity if we assume not that it was the presiding and universal sentiment of a monolithic political class for an entire generation, but rather that it was a tool that could be deployed in various situations on behalf of the interests of different groups. The conflict of the Venetian Interdict is approached here both through the long disputes over the Jesuits' competition with the University of Padua, and through the lived, confused, and improvised experience of Sagredo based in the Friulian fortress of Palmanova. Such experiences, I show, directly informed Sagredo's brilliant, funny, and cruel anti-Jesuit epistolary hoax of 1608 and his more general approach to documentary culture, whether dealing with diplomatic disputes or astronomical conflicts.

Venice's empire, emporia, and interests lay in the eastern Mediterranean. Italocentric accounts of both the Renaissance and the Scientific Revolution have traditionally cast Europe as the natural zone of innovation; this model still underpins many studies of imperial science. But both "Europe" and "Italy" are historiographically weak terms for accurately conceptualizing early modern space. As Braudel showed, in terms of both time traveled and cultural uniformity, Venice was probably closer to Istanbul than to London.[8] When considering Venetian natural philosophy, we should map the geographical zones employed in experimentation, rather than assume an innate Europeanness to science.[9] Sagredo again offers a good opportunity to study the dynamics of Venetian natural philosophy: while stationed in Syria, he corresponded with Isfahan and Goa; his network of correspondents, both direct and intercepted, extended across the Indian Ocean. Only one letter was ever sent to London; the furthest west Sagredo ventured was Marseilles. Galileo's tidal proof for the rotation of the earth is compiled exclusively from Adriatic data. Universal macrocosms were easily extrapolated from extremely local microcosms; the solution to the problem of deduction was geopolitical.

Another recent historiographical trend is to reconstruct the relations between the nascent ways of knowing of the natural philosopher and those ap-
pertaining to other identities. The case of Sagredo, who himself overstepped
the limits of professional decorum by purloining correspondence from travelers and messengers in Aleppo, is revealing, as it clearly demonstrates that
the realities of these relationships will be found not only in guides and manu-
als, but also by following different people's practices in the field.

Moreover, the proximity of the case of the dodgy diplomat to the supposedly pristine eponymous messenger guaranteeing the authenticity of the *Sidereus nuncius*'s observations should make us reconsider what was meant, and
understood, by Galileo's classic title. In what follows, new archival evidence
is presented showing Galileo's late alternatives for the book's title; and this
may be productively read alongside other contemporary cases of distant envoys with suspicious credentials, such as Xwāje Ṣafar, an Armenian merchant
sent from Shah Abbas in Persia to Venice (stopping off in Aleppo on the
way, where his mailbag was rifled by Sagredo), whose received identity was
based on the difficult task of translating and authenticating his diplomatic
documents.

What emerges from the following chapters will be, I hope, surprising and
disconcerting, provocative, suggestive, and even entertaining. Their historiographical methodologies have been improvised and assembled as needed,
rather than borrowed or preformulated, and while the results may be at times
inelegant, they are offered as potential paths out of the confines of what some
practitioners already describe as a self-regarding and hyperprofessionalized
discipline. Inevitably, and for better or worse, to some extent we resemble
our subjects. Sagredo tended to end up tangled in disputes, laughing. He may
have something to teach us.

The Generation and Dissolution of Images

To contemporary readers of Galileo's 1632 *Dialogue upon the Two Main Systems of the World*, one of its most stunning successes was the realism with which it depicted its protagonists, especially the Venetian patrician in whose palace the discussion took place, Gianfrancesco Sagredo. As Galileo and Sagredo's mutual friend Fulgenzio Micanzio put it, "My God, with what dignity you brought that worthy character Sagredo to life! God save me, but I think I hear him speaking!"[1] Galileo claimed in the book's introduction that one reason for writing it was to memorialize his love for his dead friends: "May these two great souls [Sagredo and Salviati], always revered in my heart, accept with favor this public monument of my undying love, and with the memory of their eloquence may they help me explain to posterity the promised speculations."[2] Sagredo and Salviati are not generic types, but manipulated memories deployed both to make the dialogues convincing and to rest as a cornucopian wreath on their graves.

Sagredo is now remembered because of Galileo; he lived posthumously. Sagredo left no published work, invented nothing, gave his name to no theory or law. The archival skeleton preserved by Venetian bureaucracy is unremarkable: born in 1571 to a noble family with its eyes on the Venetian dogate, he studied with Galileo at Padua in the late 1590s. In 1605 he was appointed treasurer at the desolate Friulian fortress of Palma, where he sat out the greatest ideological conflict of the century, the Venetian Interdict. From 1608 to 1611 he served as consul in Aleppo, Syria. After his return he worked for a while in the equivalent of the Venetian Ministry of Trade; aged forty-nine and studiously avoiding doctors, he died of an excess of catarrh. Why, then, bother with him?

Part of the answer was provided by Galileo: Sagredo was an interlocutor, a Socratic midwife, the charming embodiment of an ideal reader whose conversion to the new science might be witnessed and emulated. In the *Dialogue* Galileo narrates Sagredo's experiences and makes them stand in for experiments. The construction of Sagredo's identity and the credibility of his accounts underpin Galileo's arguments concerning motion on a ship and the speed of wind, for example.[3] Galileo depicts Sagredo as the ideal proponent of both common sense and shared experience. In order to do this, he deploys a carefully constructed image of Sagredo as he might have been during the period 1608 to 1611, when he served the Venetian Republic as consul in Aleppo. Galileo's use of Sagredo's image suggests an intriguing idealized exchange between Venetian state power and systems of credit in establishing the new sciences.

But Sagredo was also an interlocutor in a deeper sense: he intervened in debates and intercepted information, he transcribed and diverted documents, he manipulated knowledge through power and insisted on the political nature of scientific practice. Sagredo's identity, as revealed in his one-sided extant correspondence with Galileo, was multiple and mutable. His contributions to natural philosophical matters were varied and complex: he negotiated with glass-makers to produce lenses for Galileo, he sent his own experimental results and descriptions of instrumental innovations to his erstwhile teacher, he read Galileo's works in manuscript and print and, while his enthusiasm tended towards sycophancy, he often disagreed with him. But he performed stranger roles, too: his zealous loyalty to Venice led him to invent his own pseudonyms in his epistolary attacks on Jesuits after their expulsion from the Veneto.[4] In the debate over the nature of sunspots between Galileo and a pseudonymous Jesuit, Sagredo not only copied and circulated Galileo's texts as they passed through Venice on their way from Florence to Augsburg and Ingolstadt, but also made his own vicious, independent attacks on Galileo's opponent "Apelles," accusing him of hypocrisy and stupidity. His letters to Galileo display the witty and informal tone punctuated with mordant acuity celebrated in the *Dialogue*. There are many other versions of Sagredo also revealed in his correspondence: his instinctive sense of political satire, his refreshingly explicit accounts of his libidinal economy, his self-conscious experimentation in finding out his tastes and desires. Of all the narratives celebrating Sagredo that Galileo might have chosen for the *Dialogue*, the corpulent diplomat seems the least real and the most formal, authoritative only because false.

The only mention made of Sagredo in early biographies of Galileo from the 1650s either reduce him to his fictionalized persona in the *Dialogue* or invent details upon no factual basis. Viviani's *Vita di Galileo* omits him from

Galileo's circle of Venetian friends, and mentions him only within the context of Galileo's written works (presumably Viviani knew something of Sagredo's anti-Jesuit activities and feared censure); Gherardini's notice imagines him a seasoned ambassador, instrumental in Galileo's appointment at Padua, and, most improbably, first meeting him in Florence (where he never went) while returning from a nonexistent ambassadorial trip to Spain or Rome in 1592, five years before Sagredo would become legally eligible even for consideration for such a position.[5] Sagredo's subsequent posthumous *fortuna* was summarized neatly by Foscarini in 1754: apart from Galileo's ventriloquism in the *Dialogue*, he said, "In Venetian books, one does not meet with a single person who even remembers that there had been in this world a Gianfrancesco Sagredo."[6]

Sagredo, though, was deeply concerned with constructing and disseminating his own image. His interventions in the sunspot debate elicited a splendid, and much-quoted, declaration of identity and interest in a famous letter to the Augsburg antiquarian Marcus Welser that Sagredo copied out and sent also to Galileo:

> I am a Venetian gentleman, nor have I ever used the name of a "man of letters"; I am fond of those that do and always look after them, and expect no advancement of my lot, nor purchase of praise or reputation from the fame of understanding philosophy and mathematics, but rather from the integrity and good administration of rulers and the government of the Republic, to which I applied myself in my youth, following the customs of my elders, all of whom have grown old and consumed themselves in this. My studies tend towards the knowledge of those things that as a Christian I owe to God, as a citizen to my country, as a noble to my house, as a member of society to my friends, and as an upstanding gentleman and true philosopher to myself. I spend my time serving God and country; being free from familial cares I devote a good part of my time to conversation, service and the satisfaction of my friends, and the rest I dedicate to comfort and to my tastes; if sometimes I give myself over to speculating on the knowledge of things, Your Lordship should not think that I would presume to compete with that subject's Professors, much less enter into a duel with them, for I do this only for the recreation of my spirit, freely investigating, unshackled from all obligations and interest, the truth of certain propositions which are to my taste.[7]

In this performative story of the self, a hierarchy of loyalty is constructed. It should be read not as a timeless statement of what it means to be Venetian, but as part of a local campaign of vindication against Jesuit mathematicians in a polemic whose origin, for Sagredo, was the massive crisis of the Venetian Interdict (1606–7) less than a decade before, which had resulted in the excommunication of the Doge and the traumatic suspension of Catholic rites throughout the Veneto. Mario Biagioli has pointed out that Sagredo divided the world into the local, independent, Republic and then everywhere else, run by Jesuits;[8] this paranoid polarization rendered Sagredo's identity-construction peculiarly visible. The loyalties that defined Sagredo (to God, to Venice, to the Sagredo family, to his friends, and to himself) were relationships that demanded constant work.

Various techniques were used to keep long-distance friendships alive during the early modern period: letter writing was an important instrument for papering over the troublesome absences in the metaphysical plenitude of the ideal Renaissance friendship. The blank (or occasionally printed) book of an *Album* (or *Liber*) *amicorum* would be filled with the devices and aphorisms of visiting acquaintances, who would read through the growing list they were joining. Books and medals disseminated the image and motto of some intellectuals: these might be displayed in a cabinet, where again the singular friendship was contextualized in a web of relations. The exchange of portraits was another way to become present permanently in the work space and heart of a friend. The Renaissance museum frequently created visual narratives, genealogies, or networks of friends, donors, or patrons. Giving a portrait as a gift not only established a virtual presence; it fulfilled the contract of a picture by granting it its ideal viewer, making explicit the implied gaze. This kind of portrait, then, does not just represent the sitter: it extends a third, social dimension through the perpetual motion of friendship. The presence of the recipient is required for the work to become whole.[9]

We know, both generally from Sagredo's correspondence with Galileo and specifically from a poem written for Sagredo, that the viewing of paintings was preferably a sociable act for him. In his 1601 *Odes*, the poet Guido Casoni describes Sagredo contemplating a *Narcissus* with his close friend Sebastiano Venier: the painting set them off on a discourse concerning the nature of love and the pitfalls of modern narcissism.[10] Genuine love, Casoni's comments imply, should reach out and embrace another, not destroy itself gazing in the mirror. Similarly, connoisseurship should be not a solitary vice but a social bond: *Narcissus*'s self-destructive spell could be broken by fertile conversation. This would restore productivity to the gaze.[11]

From about 1599, Galileo frequently stayed with Sagredo in Venice when he visited from Padua. The last time the two friends saw each other was in 1608, before Galileo's rise to astronomical fame and infamy, but he wrote to Fulgenzio Micanzio in 1636, nearly thirty years after he had last seen his friend, that he regarded him as his "Idol."[12] This is more than a figure of speech: in the summer of 1619, less than a year before Sagredo died, he and

Galileo exchanged portraits.[13] Sagredo's letters from 1618 and 1619 are full of references to the initial composition of his portrait by Leandro Bassano, its slow and troubled execution by his brother Gerolamo, and its eventual dispatch by Sagredo. The painting hung in Galileo's living room while he wrote his two great final works, the 1632 *Dialogue* and the 1638 *Discourses*, both of which featured Sagredo. Upon Galileo's death, "six portraits of his friends" were listed among his possessions, the only works of art he owned apart from two landscapes in the *salotto*.[14] Despite the best efforts of the great nineteenth-century Galileo scholar Antonio Favaro to locate the portrait of Sagredo, it proved impossible to trace. The aura of this singular friendship seemed to have dissipated forever. The trail of this Galileian "Rosebud" did not die with its owner, however; its disappearance is a slow affair, lasting over a century.

All we knew of the subsequent history of the portrait is contained within a footnote to Marco Foscarini's eighteenth-century guide to Venetian literary culture:

> Galileo kept two portraits, just as he desired—one, of his student Viviani, the other of Sagredo. These are still in the possession of his heirs, and we have a copy of the Sagredo via [Dr.] Antonio Cocchi, in whom gentility of manners competes with solid science and choice erudition. The copy of the portrait was made from a life-size painting in the house of the Pansavini [*sic*, Panzaninini], nephews and heirs of Vincenzo Viviani, who was Galileo's last student, who bought from his heirs all his books, writings, pictures, instruments and learned things. After the death of Viviani, Galileo's belongings (along with a lot of other things) passed into the hands of the Abbott Jacopo Pansavini [*sic*, Panzanini], whom Dr. Cocchi heard say a thousand times that the portrait was of Sagredo, introduced into the *Dialogues* of Galileo. The portrait was next to one of Galileo himself of the same dimensions. This tradition was kept alive after the death of the Abbott, and is still going. And while there is no inscription in the painting itself, the costume is of our [i.e., Venetian] gentlemen.[15]

The only other trace of (perhaps) this painting in Florence is an undated inventory, probably from the eighteenth century, describing the portraits of famous men in the Capponi collection. In the company of Copernicus, Tycho, and Kepler (but also Savanorola, Erasmus, and Sarpi) are several portraits of Galileo at various ages, one of Salviati, and one of Sagredo. It is far from clear whether this series consists of later copies or originals, or when it was dispersed.[16]

Less than one hundred and fifty years later, Antonio Favaro could find no trace of this painting or its copy, and supposed them destroyed or, almost worse, misattributed. The lack of inscription on the original noted by

Foscarini situated the identity of the sitter in a fragile oral tradition; with this tradition lost, a firm identification would seem unlikely. But there was a small clue that Favaro seems to have overlooked: Sagredo's letters to Galileo describe in detail not only the painting of his portrait in 1619 by Gerolamo Bassano—based on a sketch by his brother, the more famous Leandro—but also, in a single reference, another portrait also by Gerolamo, executed in 1612. While the trail to the 1619 portrait seemed corpse-cold, there was still a chance that the 1612 painting might yield further clues.[17] A combination of scholarship, serendipity, and developments in the digital humanities helped me solve the mystery of the missing portraits and identify not one but three portraits of Sagredo, including that belonging to Galileo.

Sagredo in Zhytomyr

The gradual digitization of back numbers of journals has strange effects on scholarship. Unless they are well indexed, many publications in journals swiftly become invisible; book reviews, especially, used to be hard to locate and impossible to skim. One of the most important changes wrought by projects such as JSTOR is that we can know again what has been forgotten, not only within the field in which we are specialized, but across fields. Here is a typical example: *The Burlington Magazine*, one of the world's leading journals for art history news, digitized its archive, back to the first volume in 1903, in the early 2000s. Suddenly, every word of every article became searchable, including book reviews. Since reading Sagredo's description of his portrait gift to Galileo and Favaro's frustrated accounts of his failure to locate the painting, I had wondered whether some trace might not become visible in the flotsam of the digital wave. In 2000 Sagredo's portraits were invisible to search engines, but in 2005 a peculiar notice appeared. A 1991 review in *The Burlington Magazine* of a 1986 catalogue of Italian paintings in Soviet museums included some unexpected finds: "Jitomir yields three remarkable attributions: a *Portrait of Michelangelo* by Jacopino del Conte, a *Portrait of Giovanni Francesco Sagredo* by Leandro Bassano and a portrait possibly by Annibale Carracci."[18] I rushed to locate the Soviet catalogue, which contained a description of the alleged Sagredo portrait by Viktoria Markova, as well as a decent reproduction. The entry referred, in turn, to an earlier description from a Ukrainian catalogue printed in 1981.[19] It had taken twenty cold years for news of the painting to reach the West, and another ten for the announcement to be noticed.[20]

The portrait (see plate 1) is in the Zhytomyr Regional Museum, Ukraine. It entered the museum in 1919, from the collection of the Shoduar family, along with the two other pieces mentioned in the review. Baron Stanislav

Shoduar (1792–1858) was a corresponding member of the Imperial Academy of Sciences in St. Petersburg, and published a prize-winning account of foreign coins in Russia based partially on his own collection in 1837.[21] The portrait of Sagredo was previously catalogued in the museum as by, or from the school of, Moretto da Brescia (Alessandro Bonvicino, c. 1498–1554) or Giovanni Battista Moroni (c. 1520–78). Since Sagredo was only born in 1570, neither of these is possible. The painting seems to have been unstudied by connoisseurs before the 1981 Ukrainian catalogue; neither was it mentioned in any discussion of Moretto or Moroni's oeuvre. Markova's new attribution to Leandro Bassano was extremely good and almost, as we shall see, right. It was made on stylistic rather than documentary evidence. On the back it has the inscription, in a seventeenth-century hand, "Giovanni Francesco Sagredo / Veneziano."[22] It is not a particularly good or interesting portrait, but its importance for this story is that I knew I had seen the sitter's face before.

Sagredo in Oxford

Before seeing the Zhytomyr portrait, I had looked carefully through the extant catalogues of Bassano paintings in the hope that some iconographic detail might jump out and identify the sitter as Sagredo. None did, but I had compiled a mental short list of possible contenders of unknown or unconvincingly identified subjects, and one of these bore a startling similarity to the Zhytomyr portrait of Sagredo. The portrait had never been displayed, but was reproduced in a catalogue: it was in storage in the Ashmolean Museum, Oxford (see plate 2).[23]

The museum had acquired the portrait in 1935, attributed, on stylistic grounds, to Leandro Bassano. Its 1977 catalogue gave it the title *Portrait of a Procurator of St. Mark*, and wondered, with scant evidence, whether it might depict the senator Paolo Nani (1552–1608). The Bassano expert Edoardo Arslan had dated the portrait to around 1590.[24] The main evidence for this was that stylistically the portrait was thought to be by Leandro, that Leandro was ennobled in 1595, and that thereafter he tended to sign his portraits with his full title; the lack of such a signature secured the early date.[25] The portrait is obviously semiofficial, continuing the tradition of Tintoretto and his school in celebrating the Venetian oligarchy. The Ashmolean catalogue pointed out, despite the title given to the painting, that the sitter's *dogale* robe was not only worn by procurators of St. Mark, but also by other holders of high office. The *dogale* could also be worn by ambassadors and consuls.[26] Sagredo had described the portrait he sent Galileo as depicting him "in consular dress." The clothes of the Ashmolean portrait independently backed up the claim on the back of the Zhytomyr portrait that this was Sagredo.[27] Is the identity of the sitter, though, indicated by other clues within the painting?

Let's move through the painting to see what we can find (see plate 3). The book the sitter is fondling and tilting towards our view, in marked difference to the two heavy printed folio tomes behind it bound in vellum, is a gilt-edged manuscript volume bound in red velvet with ornate metal clasps and bosses. Protruding from the top is a red silk cord, on which is clamped a lead seal. This book is a Ducal Commission [*Commissione Dogale*] or prized illuminated volume of official orders, usually given to podestà, captains, and provveditori (who wore different robes), and sometimes to ambassadors and consuls, but not to procurators of St. Mark. Sagredo's own *Commissione* is lost, but the book nicely supplements the information supplied by the robe. The combination of robe and book in fact makes it impossible for the sitter to be a procurator, and means instead that he must be either an ambassador or a consul.[28]

The sitter's right hand points towards the sumptuous carpet on the table. The gesture is almost a caress, as though both the visual and tactile qualities of the piece are being experienced and offered to the viewer. At first glance this might look like a generic Renaissance Oriental carpet, but expert analysis of the texture and design of the fabric shows that it is, in fact, an extraordinarily rare and precious silk tapestry kilim. The soft fold at the corner and the floral medallion design indicate that the carpet was produced in Persia, probably in Isfahan, Kashan, or Yazd, under the patronage of Shah Abbas the Great.[29] Four such objects entered Europe in 1602 when King Sigismund III Wasa of Poland sent an Armenian merchant, Sefer Muratowicz, to Kashan to have tapestries made with the royal coat of arms incorporated.[30] Another silk kilim which probably shares the Polish provenance is now in the Residenz Schatzkammer in Munich; its border design is almost identical to that depicted in the Ashmolean portrait.[31] This is probably the earliest pictorial depiction of such a rug in European art.

Again, this evidence backs up the claim on the back of the Zhytomyr painting that the sitter is Sagredo. While consul in Aleppo, from 1608 to 1611, Sagredo corresponded with Shah Abbas the Great of Persia.[32] After Sagredo's return to Venice, he sent Abbas not only letters, but also his entire collection of devices manufactured by his instrument maker Spontino. In return, the shah promised to send him a rug. Sagredo complained to Galileo in 1612 that the rug was only worth a third of the amount he had spent on his own presents, and was still in Persia.[33] While we never receive positive confirmation that it arrived, it seems likely that this is the rare rug depicted.[34] As this may well have been the only Persian kilim in private hands in Venice at the

3077 · ROMA · Ritratto d'incognito · Tintoretto · Palazzo Doria · Appartamento privato · Anderson

FIGURE 1.1. A lost portrait of Sagredo, photographed by Domenico Anderson around 1890. Girolamo and Leandro Bassano, oil on canvas, formerly in Doria Pamphilj Palace, Rome, 1612–13. Alinari Archives-Anderson Archive, Florence.

time, its inclusion in the portrait makes the identification of Sagredo highly probable.

In fact, another Sagredo portrait has also recently emerged (figure 1.1), preserved only in a photographic negative, as the original is now lost.[35] It also depicts a Persian kilim, though this differs slightly from the rug in the Ashmolean portrait. Sagredo seems younger than in the 1619 Ashmolean portrait, and this, rather than the Zhytomyr portrait, may in fact be the 1612 portrait mentioned in Sagredo's correspondence. The sitter's face, while more roughly painted or poorly conserved than that of the Zhytomyr portrait, seems older; the costume is clearly closer to the Zhytomyr portrait, while the painting's composition, with its awkward gesture, forms a bridge between them.

A further detail in the Ashmolean portrait that demands attention is the window scene. We are obviously not looking at a view of Venice, but neither is it immediately clear what the scene is meant to represent. On the extreme left is a bridge and some kind of rotunda, to the right of Sagredo's shoulder what seems to be a series of churches with spires, and next to them a fantastic obelisk.[36] From the right, two galleys approach the most prominent building of the composition, a domed tower on a square jetty.

This building provides the key to understanding the portrait. It closely resembles contemporary Venetian idealized images of the destroyed Pharos of Alexandria, which had a domed mosque, the original minaret, erected on its roof around 875.[37] The identification is unmistakable, and no other building is depicted in this way in early modern art. But what is the Pharos doing in the background of a portrait of Sagredo? What did the building, already no longer extant, mean to his contemporaries?

The Pharos, ancient lighthouse of Alexandria and one of the standard Seven Wonders, had taken on a variety of meanings over its long history: it had variously been understood as a symbol of divine light, of political power, of the rise and fall of empire, and of superstition battling reason. More specifically, it stood, via a popular and anachronistic conflation of the cultural program of Ptolemaic Alexandria and the work of Claudius Ptolemy, as the fixed point from which the world and the universe had first been measured. One of the most popular editions of Ptolemy's *Geography*, published in Venice in 1596 and edited by Magini, transformed the publisher's mark on the title page of the book from their standard crenellated tower, with a motto of "God is my tower and strength," to a minareted Pharos.[38] Magini's account of Alexandria recalls that it was once the capital of the world, surpassed only by Rome, but it makes no mention of the ruined Pharos, describing instead the current state of the world.[39]

As Eileen Reeves has shown, textual references to the Pharos in the 1580s and 1590s show a renewed and quite specific interest in the lighthouse by contemporary Venetians as a catoptric device which supposedly magnified far-off images with mirrors: not just a building for projecting light, but an imperial telescope.[40] Reeves shows that Giambattista della Porta was the first to propose the Pharos as a candidate for the first telescope, which he calls

"Ptolemy's mirror"[41] (before the telescope had been invented), and she demonstrates that the strength of this model may actually have delayed Galileo
in his attempts to replicate reports of telescopic effects, as he concentrated
his efforts on constructing a concave mirror telescope rather than one using
lenses.

No evidence exists concerning the process by which the iconography of
the Ashmolean portrait was established. We do not know whether it was decided by the artists, the sitters, or an advisor.[42] The social dynamics that produced the Ashmolean portrait and its relationship to the earlier works are,
however, discernible from Sagredo's correspondence with Galileo.

In the Basano *Bottega*

In October 1618, Leandro Bassano and Gianfrancesco Sagredo went on holiday, along with their prostitutes. The "cavaliere" (as Sagredo usually referred
to Leandro, the only member of the family to be ennobled) was notoriously
melancholic, and the trip seemed designed to settle his fragile nerves.[43] The
task of entertaining Leandro prevented Sagredo from completing his own
mission for the trip, to test a series of lenses and telescopes. But it was also a
working holiday for Leandro: Sagredo had him sketch his prostitute, his portrait, and plates of truffles and perch. He also secured a promise from Leandro to paint Sagredo's favorite dog, Arno, a gift from Galileo.[44] Shortly after
this trip, Sagredo referred to a portrait gift in his letters to Galileo, or rather
to two paintings, an original (by Leandro) and a copy (by Gerolamo, but to
be given the final touch by Leandro). The copy was destined for Galileo to
keep; the original perhaps to serve as a model for further copies by Florentine
painters and then returned. In exchange, Sagredo asked for a portrait of Galileo "done by one of your most famous painters so that the pleasure received
by seeing your image will be joined by that which I feel for the beauty of the
painting."[45] A week later, Sagredo began to realize just how popular Leandro
had become as a Venetian portraitist: a sitter needed "the patience of Job" to
receive his painting.[46] In March 1619, Sagredo reported to Galileo that, due
to domestic disputes, the portrait was not finished, although the head had
been done "quite well." At the end of the month the head was finished, but
Sagredo had little hope of ever seeing the clothes done. He wondered whether
to have Leandro's brother Gerolamo make a copy of Leandro's version of the
head and finish it in his consular uniform, similar to a portrait of Sagredo
that Gerolamo had made seven years previously (i.e., in 1612, after Sagredo's
return from Aleppo in 1611).[47] Six weeks later, in May 1619, Gerolamo had fin-

ished the copy, dressing Sagredo in the clothes he had worn in Syria. Sagredo thought the portrait had "something new and majestic about it."[48] At the end of May, however, Sagredo still had not received the painting, as Gerolamo wanted to compare it to his own earlier portrait rather than to Leandro's newer preparatory sketch.[49] On 7 June, Sagredo received the portrait from Gerolamo and sent it on to Galileo;[50] he acknowledged receipt of Galileo's thanks a month later.[51]

Gerolamo's claim to both paintings is augmented by references in his recently published inventory. Drawn up in 1621 and comprising more than seven hundred paintings and drawings, it describes works found in Gerolamo's house upon his death. At number 52 there is a somewhat generic description of a portrait on the back of a copy of a Titian Madonna: "A portrait of a Sagredo finished behind the said Madonna by the late Signor Geronimo [sic]." Another reference is more precise: number 408 is "A portrait of Signor Gianfrancesco Sagredo finished by the hand of Signor Geronimo." In the evaluation of the collection feature two versions (or perhaps a pair) identified only as "Sagredo and Apollo."[52]

This documentary and iconographic trail to and from the Ashmolean portrait has done more than restore a name to a face. It is only by restoring the integrity of the image from its apparent fragments—now identifiable only through textile and book history, iconographic research and the serendipity of lost clues—that we might begin to understand its force and meaning. The paintings open a window onto the processes of mythologization by which a new scientific instrument becomes legitimized. The telescope is claimed both for antiquity and for Venice; Venetian diplomacy and history are presented as the systems within which telescopy makes best sense. The identification of the sitter in the Ashmolean portrait as Sagredo has been generally accepted, and the painting moved from storage to public display.[53] Some of the power has been restored to Galileo's "Idol."

Flight

Galileo himself initiated the process whereby this visual iconography was replaced by a literary representation. Posthumously, Sagredo's presence paradoxically both fragmented and augmented: his "immortalization" in the *Dialogue* and *Discourses*, written under the owlish gaze of the Ashmolean portrait, demonstrates the efficacy of early modern beliefs in the ability of representation to render what is absent present. Various other cosmopolitan myths briefly flickered around his memory: in 1647 manuscripts circulated in

Paris describing the Italian mathematician Tito Livio Burattini's project to create a flying machine for the Polish court. Burattini claimed that Sagredo was his inspiration:

> From my youth I have applied myself to this invention [flying], as it seemed strange to me that human ingenuity might not be able with its wisdom to attain those things granted naturally to other animals inferior to us, considering too that it has not been granted us by nature to live in the water, and yet by studying we learn to swim, and not just on the surface, but for a long while we go under the water, an element contrary to our breathing.
>
> In the end, having long considered these wonderful things, I resolved to attempt the mode already done by the most illustrious Signor Giovanni Francesco Sagredo, Venetian nobleman and most noble mathematician, consecrated to immortality first by his own works and then by those of Signor Galileo Galilei, noble Florentine and another Archimedes of our times. It was said to me that this gentleman had laboured hard to discover this invention upon which he had travailed for a long time, and in the end, taking a falcon and removing the feathers, and then carefully observing the proportion these had with the body both in respect to weight and size and then he made wings for his body with these proportions which he attached to his body with such artifice that throwing himself from some height he reached the ground without doing himself any harm and distanced himself many yards from the foot of the high place, but further than this he could not go by any means.[54]

The fact that Sagredo featured in the very book that would analyze and reformulate the problem of the strength of materials and scale and disprove the possibility of flying with hawk wings, Galileo's 1638 *Discourses*, should make us wonder whether Burattini was really as familiar with Galileo's work as he claimed.[55] Burattini probably grafted Sagredo into the story in order to give his own analysis of, and proposals for, machine flight more credibility. Galileian mechanics is made to go beyond its original program in order to solve what, for many, was the defining challenge of modernity: flight. Sagredo represents, in Burattini's account, the limits of a simply analogic mechanics: correct solutions had to be grounded on a thorough understanding of the nature of machines, not animals. However many times Sagredo attempted his experiment, we are told, "Further than this he could not go by any means."

The image of stately, plump Sagredo as flying philosopher has something farcical about it. But tracts such as Friedrich Hermann Flayder's *De Arte Volandi* (1627) show that the early modern flight program was epistemological as much as practical. If the angel were going to fly from St. Mark's, it would need the new philosophy to guide it. Even the most avid proponents of human flight seem always to be talking about epistemology: Galileo used the image of the philosopher eagle, soaring in solitude over the chattering, defecating starlings, in *Il Saggiatore*; the story of a volant and savant Sagredo seems designed to situate philosophical experimentation within the Venetian polis and patriciate.

Sagredo died on 5 March 1620, and his final confession, if he made it, was probably heard by Paolo Sarpi;[56] he left his estate to his sole surviving brother, Zaccaria.[57] Zaccaria gave away most of Gianfrancesco's possessions to his Venetian friends, including his magnets, lenses, compasses, astrolabes, and other instruments. This was so that his children "would not fill their heads with things of no profit."[58] He threw away as "extremely dull and improper to my profession" his brother's carpentry tools.[59] The only instrument he kept was a fine balance by Gianfrancesco's instrument maker Spontino, which he offered to Galileo, along with another snapshot of the friend's tastes: an inventory of Gianfrancesco's painting collection, intriguingly dominated by the food-preservation technique of the gastronomic still life.[60] There is no clue of what happened to Sagredo's personal archive, which probably contained at least a hundred letters from Galileo. There is no evidence to suggest that it was consciously destroyed by Zaccaria because of its content. Glimpses of potential fragments are available in later inventories. Perhaps they may still resurface.[61]

While Zaccaria dismantled the machinery of his brother's relics, lest they whir on into the minds of the next generation,[62] Galileo gathered up his literary fragments and stitched them together as the interlocutor of the *Dialogue* and *Discourses*. Even this rhetorical replication of the object of friendship could not secure the power of the idol, however; within two generations, the separation of written material from painted images and artificial machines into archives, galleries, and museums had broken the framework of meaning that could secure the permanent presence of authentic identity. Sagredo lived on only as a semi-fictional persona in a troublesome work of natural philosophy. The identity of the Ashmolean picture went from being a fact, to a conversation piece, to a rumor, and then became a memory which was soon forgotten. The Zhytomyr portrait drifted into a context where its identity was known but meaningless. The Roman portrait disappeared, leaving only a nineteenth-century misattribution. Deepening divisions between academic disciplines separated the fragments further. Their rediscovery and reidentification do not restore aura to the image, but might go some way toward allowing us to understand the affective dynamics that circulated in the production of early modern natural philosophy, and to understand the techniques and processes by which historical identity is formed and dissolved.

Becoming a "Great Magneticall Man"

At some point, most probably during his Paduan period, Galileo cast Sagredo's horoscope.[1] At this time, Sagredo was involved in trafficking horoscopes: he passed on one of Galileo's "nativities" to an unnamed recipient in 1602 and supplied Galileo with the precise birth time of an unfortunate member of the Morosini family who had fallen to his death from a bell tower, presumably so that Galileo might check the accuracy of his predictions against a recent case whose length of life (one of the categories Galileo calculated) was already known. In 1605, Sagredo illegally supplied the Friulian noble Giovanni di Strassoldo with the birth date of the new pope, Paul V, for him to cast his horoscope. This was risky information, though it need not imply machinating malice, as the source of another of Strassoldo's informants was Pompeo Caimo, doctor to Cardinal Montalto, a supporter of Camillo Borghese for the papacy.[2] In many different ways, astrology was intricately tied to early modern politics.

Galileo's astrological judgement of Sagredo's character and life strikes us as inherently biased, based as it is on firsthand knowledge of the subject, but that should not make us consider it insincere, or, on its own terms, inaccurate.

> Venus being alone is an auspicious signifier of character, free from the aspects of malefics and, being illuminated by the benign rays of Jupiter and Mercury in Sextile, determines the most honest, praiseworthy and best of characters, and makes him kind, happy, merry, beneficient, pacific, sociable pleasure-loving, a lover of God, and impatient of troubles.[3]

We do not know whether Galileo delivered the horoscope to his friend, used it to test Sagredo, or to test his astrology. Given Sagredo's own interest in astrology, we might assume that he would have taken its analysis seriously. In this sense astrology was not only accurate; it was also efficacious, a self-fulfilling prophecy.

Sagredo was born into an identity that was largely already scripted, both astrologically and socially. His family, like many others in Venice, exercised political and economic power and participated in the state's endogamous rituals of patrimony preservation in order to maintain its own status.[4] Against a backdrop darkened by rhetorics of inexorable imperial decline, which were only partly based on emerging contemporary statistics, Sagredo's choices for whom he might become were stark. Named after his great-grandfather, who had been a high-ranking official, a procurator of St. Mark's, he was the fourth of six children. His family traced its roots back to the fifth century. The origin of their name, they liked to claim, was "secret" [segreto], making them excellent functionaries. The ideal of information control implied in this etymology dictated much of Gianfrancesco's life. The legacy of public service, business, or reproduction would be bestowed on Bernardo, the eldest brother; Gianfrancesco's role would be in minor public service, and he would probably neither marry nor have legitimate children. A series of historical factors and accidents saved Gianfrancesco from his script, and allowed him to invent himself in various and surprising ways.

The most important of these was the emergence of a new political force within the political class in the 1570s, known even forty years later as the *Giovani* (youngsters). It was this group that exercised its fullest power under doge Leonardo Donà during the Interdict controversy in 1606–7. Two related conflicts between the *Giovani* and their rivals granted Sagredo opportunities he might not have enjoyed in another generation: the dispute between the University and the Jesuit College in Padua, and the Venetian Interdict. His participation as a supporter of the *Giovani*, to which his father also adhered, both initiated him into the mysteries of documentary management and positioned him tactically in polemical communities.

The University in Crisis

The first of these conflicts, the crisis over the Paduan educational establishment, emerged in the early 1590s just as Galileo arrived as professor of mathematics. It culminated in the closure of the Jesuit College, which was perceived by the University as launching a threat, in 1591. Despite the fact that the Jesuit College had been in existence since the 1540s, largely in harmony with Padua's much older and more established University, growing tensions between Venice and Rome eventually erupted in a dazzling theatre of obscenity and violence. The story has been told many times, usually with a distinct bias toward Venetian interest, as part of a narrative of the inevitable emergence of modern notions of institutionalized intellectual autonomy.[5] But as historians we should resist favoring the winners, and rather pay attention to the motives and processes of conflict.

The celebrated notion of "Patavina Libertas" or Paduan Liberty should be understood historically as an elaboration on the medieval term signifying freedom from certain fiscal responsibilities, rather than the modern notion of academic freedom.[6] While we must be alert to such anachronisms, contemporaries did praise Padua's "liberty" in terms that sound distinctly modern: there is no doubt that the University and related intellectual circles were famous, or infamous, for pushing the limits of what was thinkable, or at least teachable, in a variety of fields (especially philosophy and natural philosophy).[7] Paolo Gualdo's celebrated *Life of Pinelli* famously eulogized the University: "Nowhere else will you find an Academy in which peace and quiet, nurse of the Muses, call equally the literati together."[8]

Paduan exceptionalism was grounded on religious tolerance, deep tradition, economic prosperity, or cosmopolitan demographics, depending on who was making the point. The notion of "liberty" was, of course, culturally circumscribed. When three candidates put themselves forward to become Galileo's successor in mathematics in 1612, for example, Lorenzo Pignoria remarked that were the Jewish candidate to win, "we'll have cabala in the lectern and a foreskin on the telescope sights."[9] Such casual anti-Semitism permeated the Republic of Letters. The denunciation and arrest of Giordano Bruno mark out another of the concept's practical limits. Paduan liberty, such as it was, might be seen as an accidental byproduct of the only possible solution to the serious business of international education in a world of religious and ideological conflict. Foreign students brought Padua much of its wealth and, while other university towns attempted to make them conform at least superficially to Catholic norms, Padua's ideology buttressed the *Giovani*'s anti-Roman stance. Yet such cases as that of the mathematician Francesco Barozzi, arrested by the Inquisition in 1587, or Cesare Cremonini, whose alleged negation of the immortality of the soul earned him repeated trials, show that the university's intellectuals did not shirk from their task of goading the Inquisition.[10] The free education offered by the Jesuit College at Padua of course threatened this education monopoly, even if, in reality, the economic threat was slight and the market probably quite distinct.

The ideal of academic liberty may thus be seen as just another temporary tool crafted and wielded by the Venetian state as part of the drama of its foreign policy played out in the domestic realm. The eruption of long-simmering tensions in 1591 between the Jesuit College and the University were not self-evident or inevitable, but a ritual orchestrated by the patriciate.

In the spring of 1591, a group of patrician students from the University, representing some of Venice's leading noble families, brought to a violent climax a long campaign of intimidation consisting of graffiti, gunfire, and vandalism. The fullest description of the incident is an anonymous denunciation to the Venetian Council of Ten heavily biased towards the Jesuits.[11] The teachers' chairs were covered with scenes of sodomy "like those one can find in the worst bars of any brothels on earth," confessed the source, a little too knowingly. The unruly aristocrats threatened the priests with guns and insulted them. Again the same night the group came out, swearing at the locals, wrecking the German Quarter, and firing their guns. They went to the Jesuit College and broke the windows, shouting words so obscene "that they seemed to have come from Hell."[12] The next morning, from inside one of their houses, they put a barrel ring in the window and mooned passersby. One of the nobles then rode naked around town; a group of them forced their way naked into church during mass.

Muir describes these events as an "adolescent prank,"[13] a peculiarly anachronistic and slight analysis of this ritualized sexual and sacrilegious violence. A similar term was used by the rectors of the University in their report to the Council of Ten, but this was clearly an attempt to protect the patricians involved. After a brief inquiry, the wealthy students were fined. A few weeks later, the rector of the *giuristi* (legal studies) in the University was gunned down in the street; the Jesuits were widely suspected of organizing the murder, but no evidence has been found to back up this assumption.[14]

The task of destroying the reputation of the popular Jesuit College fell largely to Galileo's senior colleague, the professor of philosophy Cesare Cremonini. Cremonini lay the foundations of his scorching diatribe against the order and its educational establishment by claiming that this "Antistudio" was a shadowy and parasitic simulacrum of the previously healthy, original, and authentic host institution. The Jesuits, he claimed, had insidiously mimicked the academic rituals of the University, overstepping the clear limits of their original educational role in an attempt to rival and usurp them. In speeches, printed rolls, and the use of a bell to call students to class, the Antistudio aped the Studio and undermined its unique authority. Despite their vows of poverty, the Jesuit priests "started by teaching Grammar to children, then accumulating riches, I don't know how, little by little, bit by bit, and ever so slowly they insinuated themselves, until they were teaching all the sciences, with the intention, I think, of making themselves Monarchs of

Knowledge in Padua . . . ."[15] Charges of absolutist epistemological monarchy
were anathema in the republican university. Precedents for the legal forced
closure of the college might be found, Cremonini argued, in the example of
the emperor Justinian, who shut down schools in Alexandria that were run
by unqualified teachers.[16]

While in Cremonini's speech the Antistudio leached students and authority from the Studio, elsewhere Cremonini was keen to establish clear lines of
transmission and analogy in order to establish the deep foundations upon
which Padua was built. In his inaugural lecture of January 1591, printed the
same year, Venice was described as a new Athens, with its wealth and values
transmigrating through time and space. Venice's destiny was precisely to become more Athenian than Athens, the center of a truly global system.[17]

Cremonini's attack on the Jesuit College was rapidly and widely diffused
in manuscript, and it elicited at least five responses from Jesuit adversaries.
Some took him to task for his shaky knowledge of the institutions of ancient
philosophy; others repudiated the genealogies of knowledge he claimed for
himself. Cremonini's opposition between a locally produced meritocracy in
the Studio and a foreign and disloyal staff in the Antistudio was shown proso-
pographically to be simply untrue; many of his claims against the Jesuits were
similarly demonstrated to be biased, exaggerated, or malicious.

Historiographically and politically, Cremonini won the debate, with his
speech justifying the process of closing the college.[18] The reason for this lies
not so much in the easy temptation to see in the student protests a precursor of late-twentieth-century mobilizations, but in a phenomenon that was
visible to Cremonini's contemporaries.[19] One of the complaints made repeatedly in the Jesuit replies to Cremonini's speech was precisely its existence as
a written text. Far from remaining within the confines of the legal procedure
that had originally commissioned it, the speech had produced copies, imita-
tions, pasquinades, and even rival authors wishing to attribute parts of it to
themselves. The Jesuits found this public information order at least as dis-
tasteful and offensive as the speech's content: they claimed, by contrast, to
keep their contributions respectfully private. Cremonini's text had already
"passed through everyone's hands [. . .] many copies had been made, so
that even in Paduan boats it is read aloud." In Padua, the Jesuits claimed,
"all the professors are competing over the speech, one making himself the
author for some Tuscan term he inserted, another for the rhetorical figures."
Reading the speech in the Senate was not enough, but then "writings based
on the speech were disseminated; other topics were added, and, to the great
dishonor [of the Jesuits] many lies were spread, orally, in writing, and even

in verses pasted on walls. And this is spreading not only in Venice and Padua, but in many other places." The Jesuits hoped that the Senate would have the good sense to distinguish between truth and rumors, knowing "how many things people say that aren't true and how many others, passing from mouth to mouth, receive such varieties of forms and growth of reputation in the remaking that from the small things based on innocent fact they grow to be monstrous, huge, and scandalous."[20] Whereas the Jesuits stressed the exclusivity and private nature of their own texts, a fact which is attested to by the single copy of each that survives in the Jesuit archives, part of Cremonini's argument was precisely the Studio's right and duty to construct and shape public opinion.[21] The Jesuits located the point of betrayal that exacerbated the relatively minor incident of student attacks into the politically grave action of closing the college as the moment when, under false pretenses, the rector of the University asked the inexperienced Marc'Antonio De Dominis to examine the documentation granting the college its status.[22] This private betrayal informing public policy was symptomatic, they claimed, of a political system that was obsessed with the control of information and opinion. Their evaluation seems accurate.

This attention to the material diffusion of anti-Jesuit literature drew on the order's long experience of polemical attacks since its very founding. The scribal publication of Cremonini's speech was not an isolated incident, but one that must have been all too common for the beleaguered priests. Copies of the speech made their way to Paris, contributing to the campaign to expel the Jesuits from France; this campaign was used in turn by the Venetians a decade later during the Interdict crisis.

Scribal publication has often been regarded merely as an index of market forces: more surviving copies of a text indicate greater popularity, and even quality, as though history were fair. This simplistic assumption has recently been challenged by a series of works, originating in literary studies and the history of the book, that seek to understand the social mechanisms deployed and constituted by scribal publication and manuscript use.[23] There is in fact some evidence to suggest that in certain circumstances the relationship between print and scribal publication added extra prestige to the "limited edition" of manuscripts; the constitution of a "public" was not always a prerequisite for establishing political, historical, or scientific facts. Conversely, the Habermasian assumption that "public opinion" and the print revolution were causally and necessarily linked must also be reexamined in the light of Filippo De Vivo's insightful analysis of the Venetian political elite's control of just such a public opinion through scribal publication.[24]

The War of Writing

The scribal publication of Cremonini's speech set an important precedent when a much larger issue required the attention of the Senate in 1606. The Interdict blizzard of manuscripts and printed tracts makes the Cremonini-Jesuit exchange seem an impotent and unimportant flurry. Rather than attempt such a vast reconstruction, I intend here to track a different route, following the interventions of a single actor, Sagredo, through these blinding squalls. By tracing the movements of an individual agent, we see fewer trees but more wood, and the paths through it, too. Case studies such as these expose the tangled undergrowth of contingencies and improvisations of our "actors," our archiving practices, and our own historical practices by which history is made. In losing the view from above, we gain a sense of the processes by which debates happen on the ground.

These are a direct result of the political and pedagogical debates from the early 1590s; in fact, as Donnelly has shown, the Jesuits made numerous attempts right up to the eve of the expulsion from the Veneto in 1606 to re-open their school: the closure of the college did not close the debate. Digging in the Jesuit archives shows that Venice's anti-Jesuit posture was far from unanimous, and that even the most vociferous opponents of the order in the Interdict crisis were in fact in direct and sympathetic contact with the order during the intervening period.

The first archival traces we possess of Sagredo as a writer, rather than as a mere subject within bureaucratic structures, date to the turn of the seventeenth century. In 1600 the Jesuit superior general, Claudio Acquaviva, began receiving requests from various young Venetian noblemen for the order to supply them with private philosophy tuition. We have only rough drafts of Acquaviva's replies, not the original letters, so reconstructing the exchange and its intentions is necessarily difficult. One might initially suspect that the letters were part of an epistolary campaign by the *Giovani* and their sons to draw the Jesuits into some kind of trap, but such a supposition is based only on a teleological account, grounded in Cremonini's rhetoric, that pits the Venetian patriciate and the Jesuits as hostile enemies, drawing inevitably towards the crisis of the Interdict and the order's expulsion from the Veneto. In reality, the period between the 1592 closure of the college and the 1606 expulsion saw the continued pedagogical presence of the Jesuits in Padua and Venice; several attempts were made, largely by the Society's supporters rather than the Generalate, to reopen the Paduan school, most notably in 1594, 1597, and 1602.[25] The 1597 request was actually signed and supported by Cremonini himself.

Despite the closure of the Paduan school, Jesuit professors remained on site, and it seems from the copies of outgoing letters in the Jesuit archives in Rome that precisely the high-level teaching campaigned against by Cremonini in 1591 was by then thriving in Venice itself. If Cremonini thought he could protect the tender hearts and minds of future Venetian oligarchs from Jesuit influence, he was clearly wrong. In August 1600, Acquaviva responded to a request voiced by Giovanni Paolo Contarini on behalf of several unnamed nobles to receive "private lessons of philosophy."[26] Another letter, to the Florentine banker Roberto Strozzi, assures Strozzi that he "should be certain of my desire to satisfy those Senators and other Distinguished Sirs," but apologizes for Acquaviva's inability to leave Rome to visit Venice.[27] A letter to Father Achille Gagliardi, who had gone to Venice after a disastrous attempt to have Bellarmine investigated for heterodoxy in Rome, mentions again the possibility of supplying a professor in logic for the instruction of Venetian nobles.[28] Gagliardi planned to set up a "Platonic Academy" in Venice, a project that never came to fruition.[29]

In November 1601, Acquaviva replied to a lost letter from Sagredo, apparently requesting the general to reconsider the proposed transfer of the Jesuit professor Paolo Valle from Venice to Milan. Acquaviva's replies to this and other similar requests show that the Venetian mission was far from lost.[30] This is the first extant letter addressed to Sagredo, and it is interesting for several reasons. First, it forces us to reconsider the standard oppositional model of conservative clerics versus free-thinking intellectuals that is still deployed by political, cultural, and intellectual historians. These ruptures did exist, and were perceived to exist, at various times but they were not a constant, even between 1591 and 1606 in Venice. Sagredo, usually seen as one of the most radical of Venetian anti-Jesuits, is here caught in respectful correspondence, along with other representatives of his class and background, with the head of the Society of Jesus. Second, far from attempting to banish the Jesuits from Venice, this group was actively soliciting instruction in logic and philosophy from them; this was precisely the higher education competition the Senate had claimed to find so objectionable. By 1604 the Jesuits were openly referring to their "College of Nobles."[31]

The overlap between the names mentioned in the Jesuit correspondence on philosophy lessons and Galileo's Paduan circle is clear: Giacomo Alvise Cornaro testified at Capra's trial that he had shown Capra a compass in 1602 and lent him one in 1605; Giovanni Paolo Contarini signed, in December 1593, as a member of the "Consilio ai Provveditori di Commun," a document that was part of the process of Galileo's application for a patent on a water pump; Roberto Strozzi acted as an intermediary in a payment Galileo made.[32] Far from standing resolute with the *Giovani* against the Jesuits from 1591 to 1606, Galileo seems to have engaged with the Paduan Jesuits not only socially but philosophically. And, more importantly, so did the *Giovani* themselves.

The thesis of longstanding oppositions between a unified Venetian radical group and the Society of Jesus must be rejected in favor of a more supple and nuanced model. This is not to deny either the existence or the perceived existence of Jesuit machinations, or that of a massive body of anti-Jesuit literature aimed at the Society since even before its inception, which gave it its initially pejorative name;[33] nor should we ignore the subsequent anti-Jesuit stance adopted by Sagredo, or by many members of Paolo Sarpi's coterie. But it would be wrong to imagine that an anti-Jesuit ideology permeated and drove all aspects of the Venetian patriciate's political, intellectual, and cultural formation for decades. Political culture is here a practice, not a position. Certainly, figures such as Cremonini and Sarpi were under constant suspicion, not only by the Venetian inquisitor and the Holy Office but also by local Jesuits: Achille Gagliardi, for example, accused Sarpi of heterodoxy in 1602. Then again, he made the same accusations against Bellarmine.

The idea that anti-Jesuitism was a reaction rather than a stance does not make it any less sincere: its mutations and deployment take on deeper meaning when they reflect political exigencies rather than automatic bias. For Galileo and Sagredo, and their relationships towards the various overlapping social institutions of differing degrees of formality in Padua and Venice, the question of anti-Jesuitism is crucial. There is ample contextual evidence to point to a tactical and local, rather than strategic and ideological, anti-Jesuitism on their part. One of the most important social groups for the discussion of natural philosophy in this period was that around Gian Vincenzo Pinelli in Padua. The group collected, exchanged, and evaluated Jesuit natural philosophy alongside more heterodox versions.[34] Pinelli had an excellent library, now dispersed, which constituted the centerpiece of Paduan intellectual life.[35] Pinelli's manuscript collection, or what is left of it, includes, for example, tracts on magnetism probably compiled at least in part by the Jesuit Leonardo Garzoni.[36] These were read and discussed not only by Sarpi and probably Galileo in the 1590s, but also later by other Jesuits such as Cabeo and Zucchi.

Ironically, one of the products of the Pinelli group was its own historiographical eclipse: Pinelli's biographer Paolo Gualdo briefly noted the presence of another erudite gathering in Venice, the Morosini group. Despite a paucity of evidence, historians' bias has granted the Morosini group a central role in the intellectual life of Venetian heterodoxy. The "nucleus" of Venetian intellectual life produced no institutional records, no documentation of meetings, no library lists or archival traces, no correspondence, but only a few scattered fragments in trial proceedings and autobiographies. This has been enough to create the myth of disinterested patrician physics.[37] The teleological desire to institutionalize science has created a ghost group; the historiographical irony here is that such a group, or at least a network, did in fact exist, centered not on the supposedly anti-Jesuit Morosini, but on the Paduan ecumenical center of exchange and brokerage, Pinelli.[38]

Tychonic Systems

Sagredo's participation in Pinelli's circle in Padua is attested by an isolated letter sent by Tycho Brahe in 1600.[39] The Pinelli circle not only became a major information-gathering organ, but also conferred social and intellectual prestige on its members and visitors. This spread beyond mere local concerns and actors. In the 1590s, Tycho had commissioned his former assistant Gellio Sasceride in Padua to produce a copy of one of his Hven instruments, which was then to be transported to Alexandria in order to check whether polar north had shifted in the centuries since Ptolemy's observations. Various rumors circulated concerning the viability of this project: Magini supposedly claimed that the Venetian Senate had earmarked three hundred crowns to send Sasceride to Alexandria, though there is no trace of such a decision within the Venetian Archives. Tycho peddled the same idea to other patrons, such as Rudolph II in Prague and Ferdinando de' Medici in Florence; he sent other emissaries over the course of the decade, Frans Tengnagel in 1599 and his eldest son Tycho (junior) in 1601. It is unclear how seriously Tengnagel took the astronomical voyage: while Sasceride managed to manufacture a large sextant with Magini, most of the emissaries' energies seem to have gone into gaining entry into erudite groups in the Italian branches of the Republic of Letters and promoting the efforts of Tycho (senior) to be included in new collections of scientific biographies, such as those being compiled by Baldi as a response to Vasari's Lives, as the central reformer of astronomy. Tengnagel and Tycho (junior), especially, seem to have come up against fierce internal resistance.[40]

While Tengnagel was ennobled by the Venetian Senate for completing his mission of delivering copies of Tycho's printed Mechanica and manuscript star catalogue, he seems not to have been able successfully to enter the Pinelli circle. The two rare copies of the manuscript star chart he carried with him were given to the doge, Marino Grimani, and to the Senate. The first of these copies was later sold, and then presented in 1633 to the English

ambassador Sir Henry Wotton, who donated it to the Bodleian Library in
Oxford. The second is in the Marciana Library. Norlind remarks that the
copy seems never to have been officially received by the Republic, and instead
went into a personal collection. It was seen by the Sienese astronomer Francesco Pifferi in the house of one of the eldest of the *Giovani*, Girolamo Diedo,
before 1604, and collated against Magini's personal copy.

In Tycho's only surviving letter to Pinelli, he tells him that "if he has not
yet been able to see them [the two books], he would easily (I think) obtain
them from Gianfrancesco Sagredo."[41] They had been presented by Tengnagel, and it is unclear why Sagredo should seem to Brahe to be the best conduit
for Pinelli to gain access, unless he was under the impression that Sagredo
had immediately borrowed the Republic's copy. Tengnagel's experience in
Padua seems to have been a major disappointment; with entry to Pinelli's
circle impossible, he instead met Galileo, and probably Sagredo. A few years
later, in 1603, a year after Tycho's death, Tengnagel recalled his disillusion-
ment with Paduan intellectuals to Magini, presumably under the impression
that Magini shared the following views:

> As for the emulators and calumnators of Tycho, it won't be my lot to grant
> them this honor, to make those obscure little men known to the Republic of
> Letters via Tycho and the splendor of his name, who were known until now
> only to the common herd and who have raved from their mere Paduan lec-
> terns or privately against anyone according to their whim. The truth may be
> neither bent nor oppressed by these owls taking refuge in the darkness (these
> Protomathematicians, I mean, especially the eminent (please God) Profes-
> sor of Mathematics [Galileo] and the other Venetian attendant his brother in
> ignorance [i.e., Sagredo]) who because of their own ignorance cannot publish
> anything of their own before they die, envy the works of others more than
> those of Hercules, and insult them with the sharp stings of words.[42]

Galileo was, even in the eyes of his most devoted supporters, peculiarly
reticent about Tycho. Tycho's cosmic proposal offered an uncomfortably
plausible compromise between the perceived failures of Ptolemaic astronomy
and the absurdities of Copernican physics, by offering a hybrid geoheliocen-
tric solution whereby the earth remained still, orbited by the sun, moon and
stars, while the other planets revolved around the sun. Tycho's paradigm was
elephantine, and the room small. Given the polemical tone of Tengnagel's letter,
it sounds as though Galileo had not kept his thoughts on the Tychonic system
to himself in his Paduan lectures. His animosity towards Tychonism continued throughout his life. Tengnagel had met Galileo in 1598 when he traveled

through Italy attempting to secure support for the Alexandrian project. Tycho wrote directly to Galileo in 1600, offering an epistolary conversation and, he hoped, conversion that would lead to both Magini and Galileo becoming the main conduits through which Tychonism might spread throughout Italy. Galileo did not respond to this offer, and shortly after, Tycho's Alexandrian project, whose patronage base he was trying to move to the Medici court, also ground to a halt, because of false rumors of Tycho's anti-Catholic machinations in the attempted expulsion of the Capuchins from Prague.[43]

Sagredo's appearance in this diatribe is surprising, and there is no other evidence of this early anti-Tychonic (and presumably, given the strong relationship with Galileo observed by Tengnagel, pro-Copernican) stance. In 1603, Tengnagel was fighting for the reputation of Tycho, whose legacy, after his exile from Hven and death in Prague in 1601, remained uncertain. Not yet adopted by the Jesuits or inserted into contemporary histories of astronomy, his system was under heavy attack, and so was his character, with polemical religious and political suspicions easily attached to him. Perhaps more important than Tengnagel's vituperative rhetoric to Magini is its stark contrast to the trust placed in Sagredo by Tycho as evidenced in a letter to Pinelli: there, Sagredo is designated the natural intermediary in Venetian intellectual life, a privileged node linking high social standing, access to political elites, and natural philosophical curiosity. Historiographically, we have tended to follow Tengnagel, seeing Sagredo as Galileo's *assecla*—a follower, sycophant, hanger-on, or, in Renaissance English, "creature." But at the turn of the century, it seems that Sagredo had already established a reputation as a trusted intermediary concerning natural philosophy in the Republic of Letters. In 1601 he was the natural choice for facilitating correspondence between Magini and Tycho; in 1605 he mediated between Magini and Strassoldo. Part of this is to do with postal systems and Venice's strategic location as the first important port south of the eastern Alps, but the role of broker also activates, manipulates, and puts on display systems of trust and credit.[44]

We are accustomed to thinking about Sagredo as a refraction of Galileo. The published sources encourage such a reading: the earliest autonomous statement is an affidavit testifying to Sagredo's instruction in the use of the military and geometric compass from around 1597; the first surviving letter we have concerns his attempt to secure a raise for Galileo from the Padua University rectors. Galileo's ventriloquizing of Sagredo in the *Dialogue* is the logical endpoint of such a reading: Sagredo exists only in his relationship with Galileo. Such a model not only depends upon an overmachinating Galileo but reduces the autonomy of the network of actors surrounding him.

We need a model of scientific collaboration in which the intermediaries and go-betweens are not just frictionless facilitators of exchange between central figures, but active agents in the constitution, evaluation, and distribution of knowledge and practice.

Failed Brokerage

Sagredo's first extant letter to Galileo, sent in September 1599 in an attempt to secure him a rise in pay to match that of peers in other institutions, attests to Sagredo's independent patronage network, in which Galileo was only the major player. Attempting to use his father's influence to exert pressure on one of the University rectors, Zaccaria Contarini, to whom he was also closely related, Sagredo exhibits to Galileo the resources he was expected to wield in such negotiations, and the resistance he met from a senior figure.[45] He told Galileo:

> I am very upset, finding myself embarrassed in a business where, having to deal with people of great authority, I see that all of my supplications are almost completely useless and without fruit. Three times I found myself in Contarini's company, but I've never been able to get a polite word from him; indeed once he told me that when they're ready to take a break from work, they'll be working on something else. I hear from other quarters that he complains about his nephews, because they do nothing but torture him on this issue, so I reckon that in his case every approach is more harmful than productive.[46]

Sagredo saw Contarini's response as pre-scripted and inauthentic, too, and attempted to outmaneuver him by writing directly to Magini in Bologna, despite the fact that the rectors had already stated that a salary war with another university was out of the question. Despite Galileo's longstanding rivalry with Magini, with its origin in the competition for the chair in mathematics in Bologna in 1588, Sagredo supported Magini, probably with money as well as information, during the long cartographic project *Italia*, and had two maps dedicated to him in the final printed edition.[47]

These early examples of Sagredo's role as go-between in epistolary exchange and professional negotiation may make him seem rather passive, but there is also evidence that already at this point, the turn of the century, he was actively constructing his own networks. The plea to the Jesuit general for Paolo Valle's retention in Venice was still a self-interested attempt to safeguard his own education, but another letter, now lost, to William Gilbert, implies a more active and expansive interest in natural philosophy, its methodologies, and its materials.

The Terrestrial Magnet

In January 1603 the Venetian Senate sent a secretary, Giancarlo Scaramelli, to London to negotiate the return of goods stolen by English pirates in the Mediterranean. Sagredo, probably acting on information passed on by his father, Nicolò—who was one of the *Cinque Savii* to sign a petition to the Venetian Senate requesting action leading to compensation for seized goods on several occasions—seized the opportunity to send a letter to Gilbert.[48] Sagredo wrote to Galileo announcing his intention to "write to the author of the magnet, to have his friendship," and asking for smart questions to ask Gilbert, as Sagredo himself had not yet finished reading the *De magnete* thoroughly.[49] The only trace we have of Sagredo's letter is contained in the published version of a letter from Gilbert to William Barlow dated 19 February 1603. Gilbert died on 30 November 1603, and this is his only surviving letter: his library and correspondence were destroyed in the Great Fire of London in 1666, along with most of the library of the Royal College of Physicians.[50] Gilbert's letter to Barlow is the only evidence we have of his knowledge of the reception of his *De magnete*, a text that would provide both Kepler and Galileo with a potential physical cause upon which to found their Copernicanism. Barlow's epistolary appendix was intended to confer Gilbert's posthumous blessing on his magnetic experimentalism and protect him from a rival accusations of plagiarism, but there is no reason to doubt the authenticity of his transcription. The section concerning Sagredo states:

> There is heere a wise learned man, a Secretary of Venice [Scaramelli], he came sent by that State, and was honourably received by her majesty, he brought me a lattin letter from a Gentle-man of Venice that is very well learned, whose name is Iohannes Franciscus Sagredus, he is a great Magneticall man, and writeth that hee hath conferred with divers learned men of Venice, and with the Readers of Padua, and reporteth wonderfull liking of my booke, you shall have a coppy of the letter . . ."[51]

There are several things to note here. First, Sagredo presented himself as a spokesperson for the Veneto magnetic intelligentsia.[52] Despite the fact that no one else is named in Gilbert's account of the letter, we may be sure that Galileo, Sarpi, and perhaps other members of the Pinelli circle were mentioned and ventriloquized. Second is the mode of delivery and social status of the letter: a state envoy, charged with negotiating restitution of goods seized in piracy and addressing the larger issue of the illegality of English piracy in the Mediterranean, conveys a message concerning the most potentially sensitive naval technology of the day—magnetism—from one maritime state to another. The third point is that the letter is received and represented as a supplement to state business. Its authority inheres not only from the social standing of the author, but from the rituals of statecraft accompanying its delivery. It is a nice example of what Simon Schaffer, following Michèle Fogel, has called an "information ceremony."[53] Were Scaramelli not to have presented his credentials correctly, or not to have been correctly trained in the protocols of Elizabeth's court, or not to have known that the last-minute postponement of his initial audience were merely a ritualized humiliation, the scientific exchange might never have taken place, and would certainly not have been imbued for Gilbert with such gravity.[54] The exchange of scientific information—indeed, the very concept of scientific information to exchange—relies on systems of exchange constructed for other purposes. The complex processes of diplomatic accreditation, relying on the close scrutiny and translation of documents, gestures, and rhetoric, is essential to the production of early modern knowledge.

Sagredo presented himself as a "great Magneticall man," one presumes, albeit speculatively, with the same play between the anthropomorphically attractive virtue of magnets and that of epistolary correspondence which would later inspire, for example, Athanasius Kircher's exchanges with the Duke August of Brunswick-Lüneberg. Magnetism had been a new topic in Padua long before Gilbert's publication arrived. Members of Pinelli's circle had been working on magnets in the 1590s, testing previous assumptions on magnetism and formulating it into a discreet and potentially coherent body of knowledge.[55] Sarpi, too, is well known for his magnetic musings in the so-called *Pensieri* of the early 1590s, despite the loss of his presumably substantial tract on the subject along with many of his other papers in a fire in 1769. For Sagredo, though, the letter to Gilbert seems to mark something of a new start in his epistolary self-fashioning, a first effort to initiate an international correspondence with a new authority, as well as a first taste of the experimental life.

Galileo, Sarpi, and Sagredo read Gilbert in 1602, and collaboratively devised and executed an instrument for measuring magnetic declination, or the divergence of a magnetic needle from true north. In September 1602, Sarpi wrote to Galileo, uncharacteristically, it seems, to ask for further elucidation of Galileo's ideas on the "inclination of the magnet with the horizon," or the phenomenon of magnetic dip, studied by Gilbert and others before him.[56] In October 1602, Sagredo passed a "declinatorio" from Galileo to Sarpi, acting, in his own words, as the instrument's "embassy." He had tried the instrument out himself, and found that while it worked, the interpretation of its results required some thought.[57] In December 1602 he again mentioned his "declinatorio" (saying that he had not yet used it) while informing Galileo of his attempt to engage Gilbert in direct correspondence.[58] But for the delicate instrument to work it had to travel, to chart the *change* in magnetic declination over space. The resulting maps, it was hoped, might help solve the perennial longitude problem by providing navigators with the actual data of global magnetic variation to compare with an idealized norm; Gilbert also hinted that such data might provide the physical explanation for the diurnal rotation of the earth and a first proof of Copernicanism.[59] Sagredo's consulship to Syria in 1608–11 provided a first systematic effort to use this instrument for the production of this new cosmogram.

Mining Interests

The right to enter epistolary networks can have a clear material as well as social basis. Sagredo had one great advantage over most of his contemporaries when it came to studying magnets: his family owned large mines in the Dolomites that actually produced the material with which to experiment and philosophize. The geographical origin of magnets is notoriously difficult to fix, but it is extremely likely that many of the specimens circulating in the Veneto would have come from the Sagredo mines.

The Sagredo family had owned land around Cadore, near Belluno, since the fifteenth century; the vast beech forests provided Venice with important timber resources and also allowed for local smelting of iron ore. Centered on Borca di Cadore, on the road between Pieve di Cadore and Cortina d'Ampezzo, the family consolidated its holdings at the turn of the seventeenth century. A printed sale act from 1601 registers the transfer of holdings from "Piero Pizzamano, son of Z[uanne] Andrea to Gianfrancesco Sagredo son of Nicolò in his name and those of his brothers" for 2,190 ducats.[60] Gianfrancesco Sagredo personally ran the administration of these holdings, although the profits were shared with his brothers. Zaccaria Sagredo, for example, was named as the recipient of payments in 1603.[61] The iron mines, where magnets were also occasionally found, were located in the next valley to the west, at Colle Santa Lucia, a few miles from Borca. The Sagredo family mines supplied about one-third of the total ore for smelting at the various Borca furnaces, but the family received a higher price for its ore than other suppliers, presumably indicating rich deposits. They employed four factors to run the mining and forestry interests, one in Borca for iron, two in Cadore for timber, and another in Venice, presumably to arrange contracts with the Arsenal, where much of their timber and iron ended up. The most important, or at least best paid, was the factor at the ironworks, who by 1615 received an annual salary of one hundred ducats. Gianfrancesco's commercial interest in magnets probably preceded his reading of Gilbert, but may well have received new impetus from it.

Sagredo's ties to his magnet-producing mines were more than pecuniary: in his will he provided one hundred ducats for twenty dowries for local women, and he invited Galileo to visit the region with him when he went on holiday with Sebastiano Venier, presumably to test magnets as well as drink Friulian wines.[62]

Magnets also fitted well into court economics: Sagredo's "Rodomonte" magnet, when armed by Galileo, became an object over which Cosimo de' Medici and Rudolph II vied, in extensive negotiations brokered by Galileo himself. Sagredo's identity as the object's owner was initially kept secret but eventually revealed by Galileo, presumably to boost the object's fluctuating value. As the magnet entered the court, it became imbued with new meanings: Galileo, long practiced in the art of emblems, mottos, and devices, turned it into an instrument embodying and naturalizing Medicean politics. Gilbert's careful protocols of experimentation, relying on the new instrument of the "terrella," had the unintended consequence of splitting magnets into two related categories: those that were symbolically magnetic, spectacularly strong, and impressively large, and those that might be used to extend the new research programs. The locales for instrumental magnets, like the precision balances developed in the second half of the sixteenth century, were socially exclusive heterotopes such as libraries and salons.[63] Unlike the balances, however, these precision instruments, designed to measure small variations in the earth's magnetic field, were made to travel, and were only productive in their relationships across space. The dream of producing a global map of magnetic declination and thereby solving the problem of longitude with a terrestrial solution would still take years to unravel, as the discovery that declination itself changes over time had not yet been made. As we shall see in chapter 5, Sagredo would make these new magnetic instruments travel further than anyone: he himself took an instrument for measuring magnetic declination with him to Syria; he also sent them to Jesuit missionaries in Goa.

Gilbert's book had turned the world into a magnet, but it took a combination of mining, diplomacy, and geopolitics to produce instruments with the potential to transform these claims into useful knowledge. For a brief period in the early seventeenth century, magnetism seemed both to provide a physical explanation to explain heliocentric gravity and to be a potential tool for solving the problem of determining longitude. Both of these hypotheses were later rejected and replaced, but not before a magnetic philosophy had temporarily bound together the local, global, and cosmic with ties of humans, instruments, and ships. Revolutions in natural philosophy are reliant on political will. Galileo's motto for Sagredo's magnet was apt: *Vim facit amor*, love makes power.[64]

Magnetism naturalized forces as incommensurable and disparate as solar rays and patronage, but its ubiquity did not preclude specificity. Mastering magnetism as a landowner, natural philosopher, courtier, or letterwriter was one new way of translating between different systems of power. The apparent exclusivity of the contexts in which Sagredo operated—Jesuit, Tychonic, Elizabethan, Pinellian—may actually be a purely historiographical artifact, but it still indicates something of the fragmentary, peripheral, and opportunistic nature of his interventions. In the next chapter we will address a similar set of vertiginously shifting contexts in order to understand the twisting tactics of Galileo.

3

Drawing Weapons

We will, for a while, leave Sagredo to his devices in Venice, turning himself from a weak patron into a potential scientific practitioner by plugging himself into an unlikely combination of networks. This chapter will shift the lens of political and institutional conflicts to look instead at Galileo in the same period. During these pre-telescopic years, Galileo enacted a similar transformation on himself in Venice. Splicing together techniques and epistemologies from various trades and disciplines, he attempted to present himself as a new Archimedes.[1]

The purposes, practitioners, and publics imagined for this new ancient science were quite different from those traditionally in place for the correct production of natural philosophy: shipyards, building sites, battlefields, and mines were supposed to provide adepts with real-life scenarios where their knowledge would make a difference. Traditionally, such a move has been viewed as a Baconian turn away from theory and toward social utility; it might be more fruitful to reconsider it as a partial fantasy, where utility is to be understood less as an actual practical goal and more as a rhetorical move. The new Archimedes allowed universities to appropriate, transform, and monopolize previously unconnected fields of practice: ballistics and buoyancy, tensile strength, and military tactics. More important, perhaps, than the supposedly internal transformations of the relationships between domains of knowledge is the related construction of communities of neo-Archimedeans. We should understand Galileo's tutorials as part of a socioepistemological process aimed not merely at making money but also at creating a uniformly disciplined virtual school of tutees. Using a carefully balanced triad of spoken, written, and instrumental media, Galileo sought to supplement his professorial activities at the university with an Archimedean agenda. Central to this activity was his continual adaptation of a calculating instrument which he called his "geometric and military compass." Designed to perform practical and abstract operations useful to the imagined careers of his students, the instrument was constantly tweaked and updated, both to render its former incarnations obsolete and to inscribe permanently the developing needs of its initiates.[2]

Galileo also made extensive use of scribal publication to control and disseminate his works on fortification, hydrostatics, and mechanics, employing from 1603 an amanuensis to sell copies to his students.[3] He continued to use this system alongside print publication for the rest of his life. His first use of print was probably the pseudonymous and coauthored *Dialogo de Cecco di Ronchitti da Bruzene in perpuosito de la stella nuova*, printed by Pasquati and published by Tozzi in Padua in 1605.[4] The work was dedicated to Antonio Querengo, a passionate exponent of the learned rusticity exemplified by Angelo Beolco, known as Ruzzante, in his plays dating from the 1520s and '30s. Rather than capturing an authentic dialect voice, then, this form of Ruzzantian literature is strongly prescripted, but works to bind together its community of practitioners. Its form and style have often been read as a critique of closed academic elites, but in reality it merely appeals to the preferences of another kind of elite, the Friulian and Veneto literary set. The dialogue itself is of limited interest, apart from containing Galileo's first reference in print to Copernicus. But it does provide an interesting precedent for Galileo's later coauthored works and his vehement attacks on opponents' pseudonymity.

The *Dialogo* was a response to a *Discorso* by Antonio Lorenzini.[5] Almost contemporaneously, another astronomical and astrological work was published, which Galileo took as a personal attack on a series of three lectures he had delivered on the new star in Padua, though its central target was also Lorenzini.[6] The *Consideratione astronomica sopra la nova et portentosa stella che nell'anno 1604 a dì 10 Ottobre apparse, con un breve giudicio delli suoi significati*[7] was by a former student of Galileo, Baldassarre Capra.

Capra's approach to the problem of making and securing new knowledge was extremely similar to that of Galileo: he gestured at welcoming the free and easy exchange of information while demolishing and ridiculing his opponents. The discursive nature of knowledge production was, he claimed, centered not on a single author but in the processes of observation, writing, reading, and responding. His own text, he maintained, was mutable, and he welcomed correction and well-meant criticism.[8] What was crucial in the observation of the new star was priority, and in this the student had clearly beaten the master. At the same time, he charged Galileo's lectures with a lack of precision. From Galileo's marginalia we know that Galileo presented this

vagueness as a show of modesty, and Capra's insistence on exactitude as ped-
antry. Capra followed Tycho, just as Galileo would, to establish the position
of the star by using observations from Germany.

While the methodologies of Galileo and Capra may have been standardized, their epistemologies and social resources differed. Galileo's Copernicanism convinced him that the phenomenon was in fact the exhalation of terrestrial vapors, located in the superlunary realm, illuminated by the sun. The
new star's stasis among the fixed stars was for Galileo potential evidence in
favor of Copernicanism, though the logic of this claim has been disputed.[9]

What is at stake initially seems to be more the location of authority. The
resources Capra deploys to secure his own position and attack Galileo are
social, but they bring with them epistemological implications. He professes
himself the student of his "dearest master the German Simon Mayr," proc-
tor of the German nation at Padua, astrologer, Protestant, and Tychonian.
Mayr had studied briefly with Tycho in the months preceding the latter's
death; the new star of 1604 offered an opportunity to claim Tycho's authority, established in the observations of the 1572 nova.[10] One of Galileo's
defenses was to compile a chronologically ordered list of the 1572 nova observers, showing just how late Tycho himself entered the field, but also that
he still emerged as the authoritative voice and center of astronomical correspondence.[11] The astronomical dispute runs contemporaneously, in retrospect, with that over the compass. Whereas the first conflict simmered and
was sublimated into Galileo and Spinelli's pseudonymous *Dialogo de Cecco
di Ronchitti*, the latter erupted into a full dispute, imposing legal processes
and severe penalties.

Compasses

From 1590 to 1607, Galileo manufactured and sold the compass, gave lessons, and published, in manuscript form only, a description of how to use
the instrument.[12] The nexus of practices around the compass was extremely
lucrative: from 1599 on, he paid his instrument maker Marcantonio Mazzoleni between twenty-five and thirty lire per compass, plus the cost of materials and board and lodging for his wife and daughter. Mazzoleni seems to
have reached a production rate of about two instruments a month, including
some luxury examples in silver.[13] He sold the compasses for between thirty-
four and fifty-six lire.[14] His main profit was generated not from the material sale of the instrument but from instruction in its use. In 1601 he sold
an instrument and lessons for 106 lire; in the next year, in two payments
for lessons alone on the compass, he made 210 lire. Alternatively, one could

buy the instrument and a manuscript copy of instructions for 60 lire. Private
tuition—largely on the compass but also on Sacrobosco's *Sphere*, the art of
military fortification, mathematics, and other subjects—brought in most of
Galileo's revenue in the years before his transfer to Florence. In 1602 he took
a total of twenty-five hundred lire; in 1603, well over three thousand.[15] In
the same period he modified the compass, expanding its operations from
the basic twelve described in the manuscript instructions to the later printed
version's thirty-two. This textual expansion may be one of the reasons that
made him decide to shift from a restricted manuscript economy to print,
as the cost of making each manuscript copy increased.[16] In 1605 he offered
to have two presentation models of the compass made in silver, to be sent
to Cosimo de' Medici.[17] This patronage move directly foreshadows his later
tactics with the telescope.[18]

Initially, when Galileo started manufacturing his compass in the late 1590s,
he attempted to superimpose several sites of authority in the same physical
space. The compasses were manufactured at his home, where he also gave his
private tuition. He constructed a monopoly over the production and use of
the instrument. Three forms of replication necessarily threatened this monopoly: the compass left its site of production when it was sold and copied
elsewhere; the manuscript instruction booklet could also be copied and circulated in systems over which Galileo had no control; and as students were
initiated into the art of using the instrument, they could also replicate their
own experience of initiation in other sites.

By modifying and improving the instrument, Galileo could attempt to re-
situate his authority over it: every modification was a form of invention. This
process, however, risked making the instrument so protean that its identity
might move from Galileo's site of production. Reinvention could also be
replicated elsewhere, and the new compasses would no longer be his. The
modes of production, distribution, and consumption of the instrument were
inherently destabilizing. Moreover, this model of manufacture (not merely
of the material instrument, but also its accompanying techniques) was labor-
intensive. Not only profit but also work hours were directly proportional to
the number of technologies sold. The monopolistic model had a natural ceil-
ing, and Galileo did not particularly enjoy teaching. These internal limits and
related external threats put the entire system into jeopardy. As the compass
increased in popularity, Galileo's monopoly became harder to control. In the
first decade of the instrument's existence, he estimated that he had produced
more than one hundred examples.[19]

Instruments were sold or occasionally given both to members of the local Venetian and Paduan elite—such as Gian Vincenzo Pinelli, Paolo Sarpi,

Benedetto Tiepolo and Federico Cornaro—and to foreign noblemen supplementing their courses at the University with private tuition. This community was extremely cosmopolitan, or at least it came from a wide variety of places. The vast majority Galileo classified in his account books as "German," "Bohemian," and "Polish." This fact should be borne in mind when we come to consider the question of Galileo's intended audience for his later published texts, especially the *Sidereus nuncius.* There were also some "English," "French," "Flamands," and others, mainly Italians, whose nationality was not mentioned, apart from rare cases where a local identity such as "Cremonese" was given.[20] Sometimes Galileo seems not to have known his students' names, making such generic entries as "seven Poles"; at other times he was exquisitely careful to record correct titles and social relations between masters and servants.

The importance of the geographical spread of the tutees lies not in some vague notion of background but in the promise that these students were making knowledge on the move: following a scientific *iter italicum* as part of a socioepistemological odyssey that would return them, they hoped, to their points of origin not only utterly transformed as good humanists, but also standardized and networked into a community. The prime product of Galileo's tutorials was its network. This was constructed less as an information-gathering system than as a mode of self-accreditation. The instruments, texts, and personnel passing through Galileo's house emerged and scattered with knowledge purchased from their teacher and the knowledge that it was shared with their peers. We might then distinguish between two linked dynamics within this network: first, a radial movement centered on Galileo and based on a shared set of educational experiences, the purchase of an instrument, and its reference manual; and second, a potential or virtual dynamic whereby this shared experience, distributed through social channels with little in common apart from their contingent convergence at Padua in the 1590s, resonated between the tutees and validated their trust in the teacher, his instrument, and each other. Both these dynamics spread from Padua to what we would now call central, northern, and eastern Europe, though some of Galileo's students went further afield.

Private lessons, especially on fortification, were given to a large number of Polish noblemen, as well as to Tycho Brahe's nephew Otte. While it is likely that students or their servants were able to make personal copies, perhaps orally, of Galileo's interrelated tracts on mechanics, cosmology, fortification, and the use of the compass, the "author-published" manuscript was also for sale. Copies of the compass manuscript made by Galileo's amanuensis, Silvestro, were sold to Stanislav Lubinski, Niccolò Beataville, Philip Landgrave of Hesse, an unnamed German gentleman, and François de Noailles in 1603 and 1604.[21]

Feeling, he claimed, his monopoly over the instrument threatened, Galileo in 1606 published a description of the compass, available at his house.[22] At this point the relationship between instrument and text shifts slightly, as the textual description moves from manuscript to print. We should be wary of overestimating the effects of the change of medium, however: a standard seventeenth-century print run rarely numbered less than about three hundred copies, but much smaller print runs were possible. The *Operazioni* was published, Galileo claimed, in an edition of only sixty copies.[23] The printed version performed two functions. First, it outstripped the supply of the instrument, at least from Galileo's workshop. The text, like the manuscript, contained no illustrations of the instrument: it just explained how it worked.[24] One could not replicate the instrument from the book. Galileo would deploy the same approach for the telescope in his *Sidereus nuncius* four years later. The second, related, function of the text was to regain control over the instrument market, which was threatened by "pirates." Galileo had no patent on the instrument—he had no legal rights over it as an inventor (indeed, it seems that his compass was merely one of the best of its kind, rather than an "invention"). But he could defend it as an author.

The court-based model Galileo sought with his silver compass and printed book radically transformed this nexus of power systems. For a one-off payment of seventy-five *lire* to the printer and sixteen and-a-half to the engraver,[25] Galileo could attempt to re-situate his monopoly of production onto a monopoly of consumption. The gift of the idealized compass and the dedication of his book to Cosimo would decrease his income from teaching (in fact, by 1609 it had declined to just four hundred lire).[26] Instrument manufacture, which supplied him with a steady profit from someone else's labor, increased slightly, as did the price of each example (to about thirty lire).[27] No information is available on the profits from the book, and there probably were none, at least for the author. Galileo took the slightly unusual step of selling the book directly from his own house, thus combining the earlier monopolies on manuscript production and tuition in a single act, but also presumably to keep track of buyers as though he were still selling manuscripts. This transition to print seems less one of attaining a new notion of authorial authority, and more an economic calculation on the advantages of substituting mechanical writing for human. Most importantly, though, this shift of relations rendered the compass a *social* instrument: the centralized authority—or rather, fiction of authority—that he had established around the compass was presented to the Medici. Instead of repeatedly tutoring paying students, Galileo offered his exclusive services to his patron. In return, the amassed experience of a decade's initiation of students into the compass's operations became the property of the Medici; a community of elite mathematical users had a stable court to authorize their practices, rather than the self-constructed and precarious site of Galileo's Paduan house; Tuscany's patronage of the arts would be disseminated across Europe.

At least that was the idea. The reality was rather more of a compromise. The Medici were not too interested in the instrument, which by Galileo's admission had been on the market already for ten years. Cosimo accepted the book's dedication, but the pair of silver compasses seems never to have been presented.

The book was printed, and copies sold swiftly. By 1610, while negotiating with the Tuscan secretary of state, Belisario Vinta, for a transfer to Tuscany in the light of his far more successful *Sidereus nuncius,* he offered to reprint the *Operazioni,* as it had sold out.[28] But this success followed only after a dispute on the book in Padua. As we have seen, Galileo printed the text in order to liberate himself from an unstable model of technological production. He attempted to reconsolidate his authority over the compass by becoming its author. The year after he printed his instruction manual, a rival had the same idea: If the monopoly over the instrument could not be maintained, then why should textual authority be any more secure? In an unexpected development, Galileo found a copy of a Latin text in his hands which bore a striking resemblance to his Italian book. The *Usus et fabrica circini cuiusdam proportionis, opera et studio*[29] was again by Baldassarre Capra.[30]

Capra's *Usus et fabrica* makes its attack more explicit than his swipes in the *Consideratione astronomica* by charging Galileo's compass instruction book itself of being guilty of plagiarism. Although this claim was ultimately dismissed by the rectors of Padua University, it opens an interesting window on contemporary notions of education and knowledge. For Capra, the technique, once learned, and the technology, once bought, belong to their owner. Galileo's attempts to continue his monopoly over a nonpatented instrument in book form were invalid. Moreover, Capra's book is not just a straight Latin translation of Galileo's: Capra made additions and changes to the text, just as instrument manufacturers modified the compass itself, and he spliced in sections taken from other works. Capra accused Galileo of copying the compass from a preexisting German model, perhaps even one invented by Tycho Brahe. These terms were echoed in the priority disputes over the invention of the telescope a few years later.

Galileo's defense took two forms: the first was oral, the second printed. In a letter to his employers at Padua, he complained that he had published his version of instructions "to block the way against those who wanted to take the credit for my labor. But this measure has not sufficed, because recently the Milanese Baldassarre Capra, transferring my book from Tuscan to Latin, leaving out a few bits and pieces, adding a few unimportant details, has printed it in the same city (Padua), and has claimed with hurtful words that I am the impudent usurper of this work."[31]

It is this sense of damaged reputation, rather than economic ruin, which Galileo stressed throughout the dispute. He chose to claim to take the affair very seriously, because it offered him a unique opportunity to consolidate his position as the instrument's author. This is not merely a process of deploying some accumulated social capital; rather, the dispute and its escalation allowed Galileo to assemble a social machine in which each part worked alongside the others to produce a new force. A single, highly placed testimony would have sufficed to establish the legal facts, but here the same testimonies are repeated from different directions to act in harmony upon the same object. The verdict is only one product of this process: its real achievement is its manifestation of an authoritative network.

Galileo mustered his considerable Venetian social resources, providing written statements from highly placed friends attesting to the fact that he had been making and teaching the compass for a decade. The list of testimonies is impressive, including Grand Duke Cosimo de' Medici and Giacomo Alvise Cornaro, Capra's sole local ally. The attestations and other documentation reproduced within the *Difesa* echo the chronologically organized list of astronomical observations taken from Tycho's careful collection of data on the 1572 nova. Just as Tycho's epistolary account of observation functions as a pseudolegal testimony of eyewitnesses attesting to the veracity of the new phenomenon, complete with full documentation of the high social standing of the observers, so the Paduan depositions by witnesses secure Galileo's authority over the compass. The larger network is manifested only virtually, but it also serves to authenticate Galileo's long-standing control over the instrument:

> I say then that it was already ten years ago that, having myself reduced the instrument to perfection, and calling it the Geometric and Military Compass, I started to let it be seen by various gentlemen, showing them its uses and giving them the Instrument and its operations set out in writing: the instrument has been so highly appreciated that from then to now, in order to satisfy many Princes and Lords of various nations, I have had to have made in this city more than one hundred, excluding those that in Urbino, Florence, and other places in Germany, were made by my orders; so that there are few places in Europe where such instruments have not been transported by my scholars.[32]

The distribution of the compass is traced by names that also signify places: the Grand Duke of Tuscany; the Duke of Mantua; Cosimo Pinelli, Duke of Acerenza and Gian Vicenzo Pinelli's cousin; Archduke Ferdinand of Austria; Landgrave Philip of Hesse; Johann Frederick of Alsace; the Polish Kristof Duke of Zbaraz; Gabriel and Johann, Counts of Thenczyn; the French Count François of Noailles. Other unnamed individuals spread the compass and its knowledge throughout Germany, France, Flanders, England, and Scotland. The compass, which joined the geometric to the military, was distributed through the military nobility of northern Europe. Despite Galileo's admission that permission to manufacture the instrument had been granted to other instrument makers, the network remained heavily centralized. Rather than publish and distribute his instruction manual through the central node of transalpine intellectual life, the Frankfurt Book Fair, he kept the small edition within the manuscript economy centred on his residence in Padua.

Galileo also had Capra summoned before his friend the theologian and natural philosopher Paolo Sarpi, perhaps the most famous Venetian intellectual of the time, to be cross-examined on the contents of the book at Franceso Molino's house in Venice. This legal proceeding repositioned Capra's authorship within a network of etiquette and propriety where Galileo was in control. Galileo concentrated on Capra's name when he later published the *Difesa*, his account of the pseudo-trial.[33] Capra "the goat" was in fact, according to Galileo, not acting alone in mounting his attack. The satanic suggestion had come from "my old adversary and jealous enemy, and not just mine, but of all mankind, whose cutting and mendacious tongue is always ready to lacerate and tear all good men."[34] Galileo's diabolical puns on the proper name made his demonization of Capra extend beyond the "crime" in question to Capra's very identity.

The cross-examination concentrated on Capra's incompetence as an author and Galileo's skill as a reader. Galileo led the defendant through a tortuous labyrinth of misprints he had discovered in his close reading of Capra's book. This analysis verged on the philological, because it proved that Capra had based his translation on a purloined manuscript of Galileo's. Capra, for example, had given an example of the compass operation for producing the square of any given number. In Capra's book, $55\frac{1}{4}^2$ equals 45. Galileo pointed out that it does not, and offered to help Capra correct his mistake, by suggesting that he had miscopied $11\frac{1}{4}$ as $55\frac{1}{4}$, "the two 1s written on a slant, so that they looked like 5s."[35] Capra,

> instead of thanking me, started to shout, "So these are the colossal errors the mathematician wants to charge me with—petty typos!" So then I, leading him

down an ever-narrower street, asked him if his mistake would disappear when 55¼ was emended to 11¼, and he enthusiastically said yes. So I said, Go on then, multiply 11¼ by itself, and show me how you get 45, because I reckon just 11 squared is 121, and then you've got to add two lots of a quarter of 11, and then a quarter of a quarter, so you get at least 126, and not 45, like you say. By now he was more tangled up than ever, so in the end, to cut him free, because he'd never have untangled himself, I had to tell him that his mistake lay in the words *multiplicetur fractio 11 1/4 in se*, which should have been: *resolvatur numerus 11 1/4 in suam fractionem, nempe in quartas, provenient 45/4*, which works, and does what the operation says it does.[36]

Galileo's dialogic attack sought to ridicule both the book and its author. When Capra capitulated, he offered to print an apology. If Galileo were concerned solely with his reputation, as he claimed, this should have satisfied him, but his desire to retain his textual monopoly over the compass led him to more drastic action. Appealing to the University rectors, he demanded that all copies of Capra's books still in Padua be seized and destroyed. A total of 483 copies had been printed; the University sequestered 440 copies at the printer's and 13 in Capra's house. This went far beyond the issuance of an apology, and approached textual "*damnatio memoriae*," the cancellation of a person from memory and history.[37] Capra does not seem to have had to compensate the printer for the lost edition. Thirty copies had already been sold or sent out to potential patrons. This leak particularly worried Galileo, he claimed, and he printed his *Difesa* as a cleanup operation. But the *Difesa* also allowed him to replicate and publicize his *Operazioni*. By manufacturing a scandal and vehemently defending himself, he not only protected his reputation but also made and disseminated it. As soon as he had printed the *Difesa* he sent it to the patron of the *Operazioni*, who had previously never heard of Capra or his book.

Surviving dedicatory copies of the *Difesa* provide us with some idea of Galileo's aims in writing the book. In addition to the copy sent to Cosimo de' Medici,[38] he sent copies to several important figures in the Tuscan court, via whom he hoped to be able to restore what he claimed was his damaged reputation and to lever his way back to Florence. One of these copies was dedicated to Cipriano Saracinelli, an experienced diplomat and long-serving Medici courtier.[39] Saracinelli was also in charge of Cosimo's education, and was a crucial bridge between Galileo and the Medici.[40] Another copy went to Silvio Piccolomini, also a trusted Medici broker, who sent in return his account of the Medici attack on Annaba on the north African coast.[41] Another Tuscan recipient was Giovanni Battista Amadori, who in that same summer of 1607 had convinced Lodovico delle Colombe that Galileo was not the real author of the pseudonymous *Considerationi di Alimberto Mauri.*[42] Delle Columbe himself received another copy.[43] Other copies went to local friends and patrons: his English student Richard Willoughby,[44] and the powerful Venetians Angelo Contarini,[45] Francesco Molino,[46] and Girolamo Capello.[47] We may safely presume that further copies, now lost, went to the other signatories of the sentence against Capra apart from Molino and Capello, Giacomo Badoer, Antonio Querini, Paolo Sarpi, and Sagredo. Given that only about two dozen copies of the original edition are now extant,[48] the existence of eight dedicatory copies, which admittedly may have a higher survival rate than noninscribed copies, indicates a potentially large network of recipients.

This network is clearly divisible into two groups, Tuscan and Venetian. The dual strategy, which prefigures that deployed for the *Sidereus nuncius* a couple of years later, was for self-promotion with the Medici and self-preservation in the Serenissima. We do not know the size of this edition, but the group of bookmen known as the "Venetian Society" did include it in its list for transalpine export. The imagined commercial audience of the book was literary rather than philosophical: the *Difesa* was advertised in the 1607 Michelmas Frankfurt Book Fair catalogue within the Venetian section of the vernacular *belles-lettres* class "Libri peregrino idiomate scripti," rather than within the class of philosophy, alongside works of biography, history, comedy, classical commentary, and letter writing.

Readers of the book fair catalogue might easily have mistaken the genre of the *Difesa*, which was emerging, as we shall see in chapter 5, just after so many other Venetian defensive tracts against Roman and Milanese adversaries. Even now, it is hard to allocate it to a specific genre; it was part legal defense, part burlesque, and part technical treatise.

The thirty copies of Capra's book that had been sold before the order to destroy the edition was issued gave Galileo the opportunity to remind his patron of his clientage. He explained to Cosimo, "More than death, I am compelled to flee from any stain, which might denigrate my honour before your purity, so I have chosen to purge and reassure myself with the world and yourself."[49] Honor is not an inner state but a vector, dependent on material transmission. In the *Operazioni* and *Difesa*, Galileo projected his reputation towards the Medici court. The establishment of reputation necessitated not only an act of self-fashioning, seen as an inner disciplining and its subsequent dissemination, but a more complex process: the transformation of one monopolistic system of manufacture into a supplementary textual model, the policing of both these systems from external threats, and the destruction of a rival's reputation and books by assembling and deploying an otherwise nonexistent community of friends and patrons. Rather than a dispute merely about plagiarism, a theft of one person's ideas by another, the trial and related publication were a way of consciously putting together a community and placing it on display. The witnesses in the trial were not spectators but an active part of the spectacle. Capra was found guilty and forced into exile not by sheer weight of objective testimony, but by its careful deployment through a system which was itself also in movement: the witnesses became a textual unit; the rectors, themselves also implicit in the system as Galileo's employers, friends, and patrons, exerted an interpretive force on their testimonies; Galileo's reputation was transformed during this movement; and the fluidity of the compass through oral, manuscript print media became fixed. The trial itself, a space of restricted access and limited effect, was turned into a narrative that replicated these movements, both echoing them back to its original actors and projecting them out to the community of the *Difesa*'s readers.

4

Interceptions

In August 1605, Sagredo was elected treasurer to the Friulian fortress of Palmanova. His father had been general there, and the fortress was seen by the
Venetians as a crucial safeguard against both Habsburg and Ottoman incur-
sions, though it never saw a siege. Built in 1593–94, it was the most sophisti-
cated manifestation of mathematical fortress design of its time, generated by
and subsequently generating a large body of analysis and critique.

Sagredo's posting, for a safe and largely dull year, was one of two occasions when he left Venice for prolonged periods. The other was his consulship in Syria from 1608 to 1611. The striking irony of both positions, dutifully
fulfilled without, it seems, burning ambition to launch a spectacular career
along traditional Venetian trajectories, is that they actually made Sagredo
miss out on what probably would have been the two most important events
of his life; he spent the Interdict in a deserted battleground drawn up by paranoid cartographers, and he missed the greatest astronomical discovery of the
early modern era while alienating himself in Aleppo. Bad news for Sagredo,
perhaps, but good for the historian: absence makes the *fondo* grow heartier.

The aim of this chapter is to explore the use and movement of texts beyond the original intentions of their makers, where a simple model of authors
and readers seems inadequate. We will look at documents that were diverted
from their intended trajectories, intercepted, stolen, and copied; at documents whose archival location tells us about their use; at documents that lied
about their own nature; and at documents that should not exist. Some of the
evidence deployed in this chapter helps us think about texts in unexpected
yet familiar places: in piazzas and on church walls, for example. The anti-
Jesuit epistolary hoax carried out by Sagredo in 1608, which I describe in the
last part of this chapter, lets us think about the deceptive capabilities of texts

and their abilities to invent their own authors, disrupt systems of trust, and bear diametrically opposed meanings for different interpretative groups.

The experience of a year in the bleakest part of Friuli was not all bad:[1] Sagredo may have been chosen for the job by nepotism, but his training in fortress design must also have been a factor. Galileo had put together a serious private course on military engineering and theory at Padua, and although we have no direct evidence that Gianfrancesco attended, it seems a reasonable assumption. Palmanova's position was chosen as a potential stronghold along Ottoman-Imperial fault lines: in October 1605 Andrea Gussoni, the provedditore generale of Palmanova and an experienced diplomat and military engineer who had been an important part of the original approval process for the design of Palmanova, wrote to the doge to warn of the proximity of war.[2] The final throes of the Ottoman-Imperial Long War were, he claimed, coming perilously close to the fortress. This may have been less an expression of reality than a justification for the expense of maintaining the expensive site in which he was so personally invested. The accounts for March 1606 include requests for an extra two thousand ducats a month to cover the heightened state of alert.[3] Gussoni also sent frequent *avvisi* to the doge, originating from cities such as Graz and Vienna, charting the gains and losses of the many forces of the Long War. Fortresses were more than defensive buildings; they were also information-gathering centers.

The border conflicts of the Long War never engulfed Palmanova. Instead, a new threat, an Interdict, emerged, this time emanating from Rome, but with the potential backing of Spanish troops. On 17 April 1606, Pope Paul V excommunicated the Venetian doge, Senate, and government. The effect in Palmanova was almost instantaneous: Gussoni wrote on 19 April to report that he had just received the news from a Venetian courier. The initial problem for Venice was to enforce discipline across its hinterland subjects, and this meant demanding loyalty from leaders of religious orders across the provinces, preferably before they received similar demands from Rome. Gussoni called the local representatives together: the vicar of the Capuchins, the prior of St. Francis, the priest of nearby Palmata, and the priest in charge of the Duomo, which was still under construction. Venice's early policy towards the threat of Interdict was to prevent its effectiveness by literally refusing its publication: for example, the notices sent out to Jesuit colleges by the general of the order were removed from the walls where they had been posted. Gussoni promised that "in this fortress, and the churches of nearby subject towns, no writing concerning this matter will be affixed, and if it is, it will be immediately taken down, before anyone can see it, and brought to me."[4]

The Interdict controversy was fought not just by ideologues but by printers and publishers, and it included attempts to control the visual environment of subjects. This involved both the removal of Roman and the posting of Venetian material. At the end of March 1606, Gussoni was sent an official Venetian document concerning the Republic's position; he replied that he had "posted the printed protest in the most conspicuous places of this fortress."[5] We tend to think of early modern publication occurring only in portable formats, having little to do with the more permanent and fixed inscriptions of the built environment, but many sites, especially urban centers, contained official and unofficial spaces for textual display.[6] Political scopic power was more than surveillance: it extended to policing the temporary inscriptions of public space to inform and discipline subjects. Indeed, the practice of posting notices, or its supplement, the oral publishing of laws, was part of the apparatus of many institutions: for example, an excommunication was effected by having its notice posted in several predetermined public places in Rome, on the doors of the Holy Office and the church of Santa Maria sopra Minerva as well as in the Campo de' Fiori. Facts are by definition public. The irregular pulse of news and announcements was measurable at cities' arteries. The power of such spaces is not fixed, but fluctuates with temporal rhythms: what is officially posted during the day may be removed, defaced, or substituted by night. Sites also emerged where anonymous criticism was unofficially tolerated, and publicly consumed, such as the statue of Pasquino near Piazza Navona, in use since the early sixteenth century as the appropriating persona of anonymous Roman pasquinades. Even officially posted notices would quickly be replaced, superseded by others to form a constantly changing textual environment. The ephemerality of these notices makes them hard to study, but diplomatic correspondence is a rich source for writing the contested history of these sites of inscription.

Gussoni's attempts to control his local clerics in and around Palmanova seemed initially successful: only the Capuchins expressed some qualms about disobeying the pope.[7] A month later, however, the Capuchins were ordered to obey the papal Interdict, and they "united, prepared to give up [their] lives." They sent an offer of heroic martyrdom to the doge,[8] but also a neat compromise: they were instructed that they could still celebrate mass if they locked the doors of the church and kept their voices quiet.[9] Such a compromise was not allowed, however, as an unnamed priest then arrived from Venice ordering them to adhere strictly to the terms of the Interdict. The offer of martyrdom was also refused, with Gussoni advising that any form of violence against the priests was out of the question and would be counterproductive, given the spiritual confusion of his soldiers and the local population.[10]

Tensions had been mounting for nearly the entire period of Sagredo's presence in Palmanova, and it is probably no coincidence that just five days after the publication of the excommunication, Sagredo sent the doge, Leonardo Donà, a report on Palmanova's fortification system.[11] Gussoni was at this point campaigning for a large budget increase, and had submitted an estimate for new building works at over sixteen thousand ducats in May.[12] Sagredo's report, dated 22 April 1606, is no longer included with the original of his letter, so we have only partial evidence as to its contents. The design of Palma, with its classic *trace italienne* star, had not been superseded by Galilean military theory: "Thank God," writes Sagredo, "for the prudent deliberations of the Senate, and for the information [*intiligenza*] and diligence [*accuratezza*] of the Generals, that were used in the construction of this fortress, and which is already well on the way towards such a degree of perfection that it is impossible to want more."[13]

The fixed star of Palmanova was built, as Deborah Howard has recently shown, through dispute and discord. Marc'Antonio Barbaro, the on-site but militarily inexperienced *provveditore*, struggled with the material impossibility of perfectly realizing the designs and models in paper and wood that had been created by the old soldier and master fortress designer Savorgnan. This was a dispute not simply between practice and theory, but between different notions of authority and expertise, complicated by long-distance noncommunication and fractured disloyalties among staff.[14] Barbaro died on 4 July 1595, probably as a result of malaria contracted during the supervision of the fortress construction; Savorgnan quickly followed him, on 14 July 1595, aged eighty-eight. Palmanova, intended as a city as well as a fortress, was apparently still barely inhabited and also defensively weak in 1600; it received brick facing in 1605, and by 1610 it had cost three million ducats.[15]

The experience of the Interdict at Palmanova shaped Sagredo's approach to documentary culture. Gussoni, the *provveditore generale* with whom Sagredo seems to have been on good terms, sent regular *avvisi* to Venice to provide both breaking news and a trail of information. There is some evidence that the Friulian outpost also initiated Sagredo into *avviso* culture, though, he had already consumed *avvisi* in Venice and been exposed to them through family members engaged in state service.[16]

The first extant example of Sagredo's active manipulation of documents dates from 27 August 1606, while he was still in Palmanova. From there, he sent a letter to a high-ranking Venetian church official, the *primocerio*, Giovanni Tiepolo, describing the contents of an *avviso* received from Rome with news relevant to Venice both domestically and internationally.[17] The first piece of news was that the publisher Giovanni Battista Ciotti had been excommunicated, but given that Ciotti's excommunication, for following the Senate's orders in omitting part of volume 5 of Suárez's *Disputationes de censuris*, had been published only on 7 September, Sagredo's possession of the information in Friuli eleven days earlier is evidence of his access to something more than the usual recycled news of the *avvisi*. The start of his letter states as much: "Because I have not been able to have the replies from Rome to copy, I will tell you only this detail, that they contain the news [*l'avviso*] of the publication of the excommunication of Ciotti, for having printed a book in a mutilated state."[18] The first phrase is striking, implying that Sagredo was repeatedly intercepting someone else's letters: usually copying them, but sometimes only having the chance to read them through. The remainder of the letter reveals little of its source, and contains only diplomatic and military news. Only one phrase shows that the letters were not originally intended for a Venetian, or rather pro-Venetian, recipient: "The Spanish armada has taken Durazzo [Durrës], news (he writes) that will not be too pleasing to the Venetians." That distancing between the voice of the Roman letter writer and Sagredo's report of it reveals some subterfuge. Sagredo's letter was sent with a *frottola* printed in Rome on the taking of Durrës two weeks previously, but it is unclear whether this text was supplied by the Roman letter writer or via another route.

We do not know whether Sagredo used his official position, his family name, or an invented identity to get hold of the Roman news. What is clear is that he saw and seized the possibility of intercepting letters, and realized that even backwaters such as Palmanova might serve as temporary information-gathering nodes. This insight directed the rest of his life.

Sir Henry Wotton's Anti-Jesuit Network

Soon we shall see how Sagredo became expert at intercepting, opening, copying, and resealing letters while stationed in Aleppo. In this earlier period at Palmanova and Venice he seems to have adopted slightly different tactics, either using his reputation to draw correspondence in, or subverting this system by inventing identities to establish fake epistolary exchanges. A useful parallel example, which may well have been known to Sagredo either directly or via their common contact Sarpi, is that of the English ambassador to Venice in this period, Sir Henry Wotton. His correspondence documents the birth of an expensive new pastime of intercepting Jesuit correspondence during precisely the same period as Sagredo's interest in *avvisi*. Wotton's espionage was not initially based on a general desire for "intelligence," but was motivated by the specific target of the Society of Jesus during the Venetian

Interdict. He initially sent to London copies of letters directed from Rome to the Jesuit Antonio Possevino in April 1606.[19] His source for these interceptions was based in Rome itself, though he swiftly expanded his network of spies and informants to include Milan, Turin, Frankfurt, and even Venice.[20] Soon he became committed "to furnish his Majesty with the knowledge of the secretest practices out of the very packets of the Jesuits themselves," regularly opening, copying, and resealing their letters.[21]

Using a vocabulary close to that of Sarpi, with whom he was in contact via his chaplain, Wotton sought to expose the hidden "political instruments" (*istrumenti politici*) of Jesuits manipulating the world.[22] Upon hearing of the assassination of Henri IV of France, for example, Wotton immediately assumed that the Jesuits were the instigators, claiming in an official speech that he would rather know Ravaillac's confessor than his confession.[23] In Wotton's analysis the Jesuits were not mere passive tools of political elites but powerful actors in their own right, whose own disputes with rival orders were the direct cause of the Interdict:

> I must humbly confess unto your Majesty that for mine own part I have here, by discourse upon the place, deeply received into myself an opinion that the Jesuits were the first moving cause of the present troubles, and not as instruments of the Papal or Spanish, but of their own greatness, either to divert the Pope (as some, expert in the Roman Court, conceive) from decision of the controversy between them and the Dominicans, to whose side he be more inclined; or that, in a troubled and active time, they might be sure, through their intelligence abroad, to carry him even from his own cardinals, and to be the masters first of his ear, and within a while of his determinations.[24]

In order to reveal the hidden tools of his enemies, he had to place his own "instruments," and it is telling that he uses the same term for his own system as theirs.

Wotton's construction of a mirror system to expose Jesuit fraud gradually began to replicate its object: in November 1606 he sent James I an intercepted discourse by the doge concerning Spain; this was precisely the kind of machination he claimed to find so immoral among the Jesuits. The efficacy of the entire Jesuit information order was threatened by its rotten Italian core, Wotton claimed. In a speech to the Venetian Collegio delivered in June 1607 and reported to the Venetian ambassador in Rome, Wotton protested a Jesuit pamphlet that rewrote the Interdict settlement in Rome's favour. The Calendar of State Papers translates his speech:

> I know the wiles of the Jesuits and how they manage affairs to their advantage. I see the news they spread abroad from Japan, from the Indies, from the

New World. That is all right for the salvation of men's souls; and if any one
has doubts he may go and see for himself and verify them. But to attempt to
obscure the affairs of Italy, patent and open to all, this rouses wonder and
stupefaction.[25]

Wotton's interceptions themselves became the subject of his intercepted
letters. This encouraged him.[26] The "trade of intercepting letters," as Wotton called it, gained momentum as it became institutionalized and received
a growing budget. The information system then started to assimilate other
epistemologies: the medium really was the message. Consider, for example,
Wotton's famous account of the publication of the *Sidereus nuncius,* where
Galileo's astronomical discoveries are described primarily as "news." The
gratifying fullness of Wotton's description of the book's contents has diverted attention from its larger context. But this is how the Galileian snippet
actually fits in to the flow of Wotton's letter: he opens with a discussion of
the resolution of a scandal concerning a reformist letter from the Calvinist Bible translator Giovanni Diodati to a French politician and religious reformer that had been intercepted and presented by the Jesuit Pierre Cotton
to Henri IV of France. Altered copies had been sent to the pope and to the
French ambassador in Venice, in an attempt to suppress reformist tendencies
in Venice.[27] Wotton then offers Salisbury a brief analysis of the political sympathies of the various organs of the Venetian political system, and requests
that he may return to England via Germany in order to make some direct
contribution to the printed propaganda supporting the conversion of Venice
to Protestantism.

After this discussion of future plans, Wotton returns to "the occurents
of the present": "the strangest piece of news (as I may justly call it) that he
hath ever yet received from any part of the world," the *Sidereus nuncius.* He
provides not only an efficient synopsis, but also an immediate appreciation
of the radically destructive message of the book for both astronomers and
astrologers. Suspending judgement on the truth of the book, he passes on to
other news: "Now to descend from those superior novelties to those below,
which do more trouble the wise men of this place. Our discourses continue
with increase, rather than otherwise, touching the secret purpose, accorded
between the French King and the Duke of Savoye, to assail the Dukedom of
Milan, whereof I am unripe to render his Majesty any farther accompt."

What is of interest here is that Galileo's discoveries are presented precisely
as if they were contained within an *avviso,* whose proper place of analysis is
within a broader information system. Wotton was perfectly capable of writing about technological inventions within a philosophical framework: there

is a letter from around the same date to Prince Henry concerning an invention for preserving gunpowder which contains explicit musings on the nature of technological innovation and use.[28] But the *Sidereus nuncius* is not presented in this way: it is a product of a new kind of "optical instrument (which both enlargeth and approximateth the object)."[29] It makes things bigger and brings them closer; it is an instrument to be understood within the domains not of philosophy but of politics. As if to prove Wotton right, the day after he publication of the *Sidereus nuncius*, Henri IV of France was killed. Better instruments might have prevented it.[30]

Post-Interdict

Sagredo returned from Palmanova to Venice in spring 1607. In November of that year he was elected consul to Syria. On 1 August 1608, Sagredo's friends threw him a farewell banquet at the Barozzi palace on the Canal Grande. He left the next day for a consulship in Aleppo, and was far from sure that he would make it back alive. Galileo, away in Tuscany tutoring Cosimo de' Medici, was unable to attend. He would never, in fact, see his closest friend again: by the time Sagredo returned, Galileo had moved permanently to Florence and, despite Sagredo's best efforts to have him resume his position at Padua, would never again enter the Veneto.

The leading Latin poet among the Venetian scanning classes, Ottavio Menini, presented a propempticon, or ode for a departing traveler, composed for the occasion. Menini was a senior and respected figure among both Friulian and Venetian literati, both as a poet and a critic. He worked his way into Venetian political elites, writing celebratory verses for various doges and cardinals and corresponding with other literary beacons such as the Benedictine Angelo Grillo, whose letters he edited to the author's dissatisfaction in 1602. During the Interdict, Menini sided closely with Sarpi. One of his tracts, the *Panegyricus serenissimi principi Leonardo Donato* (1607), was even banned in Rome. He failed, however, to secure a permanent position within Venice after the Interdict, and in 1608 he entered into secret negotiations with Rome for a transfer. Potential patrons there balked at his salary request, and when the dealings were discovered in Venice, he lost his pension there, too. He stayed on in Padua, his circle of correspondents overlapping with that of Sarpi, who eventually despised him. In 1610 he reopened Roman negotiations, a process which continued even after his appointment as lecturer in Latin at Padua in 1614 at the age of nearly seventy.[31]

Menini's ambiguous and ambivalent relationship with Venice is crucial to understanding not just the subject matter of his compositions, but his decisions to publish, both scribally and in print. Many of his occasional pieces were assembled in 1613, probably as part of his Paduan job application, and printed in Venice under the title *Carmina*. The collection includes poems dedicated to many members of Doge Donà and Sarpi's circle, such as Sebastiano Venier, Antonio Querini, and the French Huguenot Jérôme Groslot de l'Isle.[32] There were also some poems designed to impress the Roman curia, but the predominant cultural policy was towards a French-Venetian alliance. A poem celebrating print, for example, makes no mention of Gutenberg but instead celebrates Robert Estienne and Aldo Manuzio.[33]

Included in this collection was the Sagredo propempticon, which previously has only been known through a single anonymous manuscript copy among Galileo's papers.[34] Unlike many of the works in the *Carmina*, the dedicatee or subject is not mentioned in the title, but only within the body of the text. There is no reference to the poem in Galileo's correspondence, and its anonymity, contents, and handwritten status make it likely that the text had an extremely restricted initial circulation. Sarpi sent a copy to his correspondent (and another object of Menini's muse) Groslot de l'Isle on 5 August (the poem itself is no longer attached to the letter), explaining some of the context that might make it intelligible. He also made a reference to it in an interview with a German protestant. We will return to both these documents.[35]

Favaro seems not to have read Sarpi's letter fully when he discovered and published the poem in 1905, even though he cites it there.[36] Although Sarpi refers to "the attached Ode by Signor Menino" in the letter, Favaro remained oblivious to its author's identity, and therefore did not see that it had already been published only five years after its initial delivery. Menini's decision in 1613 to publish what must have been considered by all in 1608 to be extremely sensitive material perhaps shows nothing more than the short shelf life of scandal, but it may also be part of his advances in his one-sided courting of Rome. The poem, then, sits astride various contexts: delivered on the eve of Sagredo's departure, it summarizes and celebrates part of his life like an premature obituary. It straddles the division between a restricted elite audience of oral and manuscript performance and a print public; it was produced by an unsuspected double agent in the midst of his schemes; and in all these breaks it marks the start of a transition from a climactic opposition of Venice against Rome to a more domesticated and conciliatory stance.

Here is the poem in translation:

> To Lightheartedness
> Goddess who shakes off cares from the breast
> And clouds from the brow, direct your swift foot here.

Come, Goddess, and approach Barozzi's happy home
And festive hearth with gladness.
A select band of leaders, the flower of the noble class,
Scepters, rods and purple,
Lacking their usual gravity, virtually oblivious of their
    surroundings,
Invite you to full cups.
The first of August truly grants that indulgence,
Than which no day is more auspicious,
Has less of a raised eyebrow,
Is less enchained by stern laws.
It is right to drink good wines on a good day:
Server of the Falernian wine, come here, boy:
Boy, offer the bowl frothing with sweet nectar
Two, three, four times and more;
Ah, may the water nymphs be far from here, and may
    foul poisons
Be banished from these tables.
Boy, offer the lute: it will give pleasure to have gripped
The ears of leading gentlemen with the lute.
Greetings, brave heroes, my guiding spirits:
May this day be fair.
May it return to you, with a hundred returning years,
Always happier and happier.
You heroes accustomed to make laws throughout great cities,
Rivals of ancient Cato,
Now indulge your spirit: this is what the fleeting hour
    demands:
You will return to great things tomorrow.
Enjoy yourself before the rest, darling of this gathering,
Sagredo: you indeed will soon
Set sail from these shores, passing over the boundless sea,
Where the beloved country sends you,
And amongst different peoples and Barbarian kingdoms
Destined to be the guardian of citizens.
May friendly stars make your journey happy
And the return joyful,
And after business transacted with a skilled hand,
After distinction gained by courage,
May they restore you, increased in wealth and distinction,
To the passionate embraces of your own people.
We here, in the meantime, will enjoy you with mindful
    memory,

And your image will not fade
From our breasts; such good looks and seductive charm
Will not ever fade.
It will be enjoyable for us to recall not only what you have
    said in earnest,
But also what is droll and fun.
But while you are away from here, who will then mock
Those grasping, greedy birds
Whom, far from here on high, the credulous crowd of Rome
Believes to be so many spotless swans?
Who will, under the false name of a rich matron,
Publish those sweet little letters?
Who will capture the inheritance hunters, and teach with
    equal art
How to defeat deceitful arts?
Oh! may you return quickly, may you return on favoring
    south winds!
For, once you have returned safely,
Your joyful companions will lift the same cups with you
And they will exalt your praises and glorious deeds to the
    heavens,
So that a fair girl will receive you back, clapping her hands
    and running to meet you,
And she will honor you like an immanent deity:
That girl, to whom Apollo has yielded his skills,
Who exudes every grace,
That girl, who was once deceived by a sacred guise
In a treacherous household,
And who recently, privy to the truth,
Revealed the trap of those who had, with deceitful hooks,
    drawn her
To themselves in her innocence
And she made it known with a remarkably charming poem.
But where am I heading? Give me wine, boy: it is always helpful
To prolong the dripping day with drink.
Oh, Sun, do not hurry: may this day last more than a year,
And may Hesperus himself not rise.[37]

The poem blends some standard toadying to the senatorial class with several specific details on Sagredo's recent activities. The "darling" (*oculus*) of
the group is sent off to a banally depicted and vaguely imagined Aleppo, even
though the city had recently been the site of a widely reported Kurdish rebel-
lion and was well known to Venetian merchants, travelers, and diplomats.

The poem devotes its energies, instead, to providing a portrait of Sagredo's legacy, as a series of future absences in Venetian cultural and political life. Sagredo's enduring presence relies on an "image" (*imago*). The group promises to keep tight its strong homosocial bond by replacing Sagredo with an effigy, but also rehearses his recent contributions to political life.

The allusions within the poem make sense only to those already initiated into its secrets. When Sarpi wrote to Groslot de l'Isle with a copy of the poem he included the following instructions:

> To explain, I'll just say that Giovan Francesco Sagredo, a noble in this republic, has carried out a solemn hoax [*solenne burla*] on the Jesuits: having made up the name of a rich widowed gentlewoman, he has got hold of a large number of replies from the hands of these wise good Fathers, full of their doctrine and arts, seeking now replies to doubts and scruples, then for advice on drawing up her will and in other ways. The intrigue has gone on for four months, two letters a week (that's how often they go from here to Ferrara). The gentleman adopted at the start the means of a "Cheatine" (as we say here), that is, woman devoted to the Jesuits, but absolutely "deCheatinised"; by such means he tricked some go-betweens of the good fathers here, who carried out the job of sending the letters. This gentleman was to leave on Saturday (as he did) for Syria, where he is going as consul. To honor him, some of us met up on Friday for dinner, where Signor Menino had the attached ode read out.[38]

This is not the only mention we have of the hoax: Sagredo had also already referred to it on a couple of occasions to Galileo. On 22 April 1608 he included this note: "From Ferrara I've had a brief reply from M. Rocco Berlinzone, who doesn't wish to dispute with my friar, excusing himself by saying that this friar seems more heretical than religious."[39] A few days later he added that "the Jesuit trial [*processo giesuitico*] runs along smoothly."[40] The way the subject is mentioned makes it clear that Galileo was already aware of the trap, and knew the moniker Berlinzone. Perhaps he and Sagredo had discussed it before Galileo's departure to Tuscany. Sagredo's copy of the epistolary exchange is lost with the rest of his papers. More strangely, Galileo's copy is also no longer extant. It was sent from Venice in 1619 and was still, more than a decade after its production, considered an important part of Sagredo's self-image. The timing of this gift coincides with a new hostility between Galileo and some Jesuits concerning the nature of comets; much of their dispute turned on conflicting ideas of correct authorship. Whether requested by Galileo or not, the 1608 hoax must have seemed a pertinent document in this evolving context. Presumably the correspondence was then destroyed, in a period when possession of such evidence would have seemed potentially incriminating.[41]

In May 1608, Sarpi sent a detailed description of the ongoing hoax to Antonio Foscarini, Venetian ambassador to France, calling it "a nice story, to
amuse you."[42] Even though the exchange was not yet over, Sarpi was ready
to start extracting lessons from it: "Now what should we say of this forged
event [*successo finto*]? Should we not draw certain conclusions? That, even
if they're exiled, they still fish in our waters and sow in our field the seeds of
that doctrine which could not be more pernicious to us." After guiding Foscarini through the hoax and its implications, Sarpi prodded the episode on
towards one of its intended audiences, mutual Protestant friends in France:
"I think those gentlemen [in Paris] will laugh when they get to hear about
this story."[43]

In 1965 Gaetano Cozzi discovered the sole surviving copy of Sagredo's
"Berlinzone" exchange, in the Marciana Library. It is not clear where the
copy came from, but it dates from the first half of the seventeenth century
and provides some evidence of limited circulation. Before entering the Marciana collection, it belonged to the Contarini family.[44] The most likely candidate for possession of an early copy was Niccolò Contarini (1553–1631). Contarini was a central actor in the Interdict controversy, as an ideologue for the
Venetian *Giovani*. He was a close friend of Antonio Querini, who is defended
in the text, as well as of Sagredo's friends Sebastiano Venier, Agostino da
Mula, Leonardo Donà, and Paolo Sarpi.[45] He may even have been involved in
the hoax's inception, granting permission for his family name to be used by
Sagredo. Cozzi considered the correspondence "the monument of Venetian
anti-Jesuitism," but it remains unpublished. The manuscript copy consists
of eighty-eight folios, made up of forty-eight letters and some supplementary material. It is divided into several sections: "From the letters of Rocco
Berlinzone Volume One," with "Volume Two" starting on f.28v and "Volume 3" on f.60v. These divisions, which do not reflect clear chronological or
logical breaks within the material, may well represent the groupings in which
the original letters, or early copies, originally circulated in installments. It
purports to be a true copy of the original, "copied from the authentic word
for word with complete fidelity and diligence." The copy's subtitle further
explains the reasons for its production: "so that from this attempt made with
this Father every man of judgment might know what to expect from the return of the Company of Jesus into the City of Venice." It was probably copied
soon after 1608, when the Jesuits were already campaigning for readmission,
though it may date to any time before 1657, when they were finally allowed
back in.

The first letter is dated 12 March 1608; the last (numbered 48), 12 July. We
know that Sagredo left for Syria on 2 August, though it was not because of

any scandal created by the hoax: his nomination for the position had already passed successfully in the Great Council by November of the previous year.[46] The manuscript also includes other types of documents: a fictitious account of the death of Cecilia Contarini, supposedly sent by a servant; a short anti-Jesuit commentary; a vicious text sent to the rector of the Jesuit college at Ferrara, Antonio Barisone; and a copy of a letter from Barisone to Sagredo from 7 March 1608 on an unrelated subject, whose original must have been elicited in order to authenticate objectively the handwriting of the Jesuit part of the correspondence.[47]

The broad content of the exchange is well known due to the accounts of Cozzi and Pavone, and usually receives a walk-on part in biographies of Galileo due to Favaro's brief discussion based on Sarpi's letter. Sagredo claimed to be a rich, old, slightly batty widow called Cecilia Contarini, with a heavy purse and even heavier conscience. Under surveillance by her meddling family, who disapproved of her devotion to the Jesuits, she insisted on absolute discretion in managing the exchange. A drop house was set up for safe delivery of the letters; those from Ferrara were to have plain seals to avoid raising suspicion. The Jesuit priest suggested that Contarini burn his letters as soon as she had read them, lest her family find them; she claimed to have done so, "only copying out certain passages lest they be forgotten."[48] In fact, as the existence of the copy itself makes clear, along with a note in the final synopsis of the exchange, "Berlinzone's letters, authenticated by other writings concerning other matters, remain with the person who procured them, along with a register of everything written by the invented widow" (ibid., f.69r).

Sagredo exploited the Jesuits' powerful position among Venice's noblewomen; his main target in this exchange was the Jesuit Antonio Possevino, who had attacked Venice under at least three pseudonyms during the Interdict crisis. Unfortunately for Sagredo/Contarini, Possevino had been transferred elsewhere, and his letters were being answered by the rector of the college at Ferrara, Antonio Barisone. Sagredo/Contarini set an elaborate trap for the Jesuit: she asked apparently naive questions about the long-term spiritual effects of the Interdict, the views of various confessors in Venice, and, above all, to whom and exactly how she should leave her considerable fortune. She gradually extracted from Barisone the kind of advice that would suitably scandalize the *Giovani* group of politicians in control of the senate. Her blithering letters in Venetian, with sentences lasting pages, were meant by Sagredo to parody a certain epistolary style and create the impression of a feeble and fearful feminine mind, but the agenda beneath the surface was stiletto-sharp. Barisone's responses read like an early source for Pascal's *Provincial Letters.*

Barisone fell completely for Cecilia's imagined and obsessively reiterated dilemmas, and failed to verify independently whether his correspondent in fact existed. In order to avoid Barisone's brave efforts to send a pro-Jesuit Venetian female friend of his to comfort her, Sagredo/Contarini invented a series of accidents and diseases that kept her in bed. She created an impenetrable defense for her cover by invoking the very climate of paranoid anti-Jesuitism that Sagredo was helping foster: in a world of spies and traitors, where even her own correspondence was not safe from her family, she suggested that they both employ pseudonyms. For herself, she chose the suitably saccharine Angelica Colomba. Antonio Barisone became Rocco Berlinzone, echoing the Nowhereland of Cockaygne called Berlinzone by Boccaccio.[49]

The correspondence not only ensnared Barisone; it also exposed a network of Jesuit sympathizers in Venice, go-betweens who were willing to break the Venetian law prohibiting contact with members of the Society. Barisone himself supplied loopholes whereby Contarini might leave her properties, goods, and money to the Society.[50] This provided documentary evidence of the Society's divisive practices and continual disregard for the well-being of the Republic. Sagredo/Contarini actually presented Barisone/Berlinzone with copies of the Venetian laws prohibiting donations to religious houses from 1536 and 1605. The copyist of the Marciana manuscript notes that at various points, where Sagredo invented documents from other fictional characters, the handwriting also changed: the production of the documents is one of their leading concerns, whether Cecilia tells Barisone that she has finally managed to obtain ruled paper to guide her undisciplined handwriting, or in the careful assemblage of fictional witnesses bearing a mixture of authentic and forged documents.

In an 1873 Venetian study celebrating the centenary of the suppression of the Society, a letter from Barisone, dated 16 July 1608, was printed documenting his knowledge of an epistolary trap. At that point in the correspondence, Sagredo's identity had not been revealed to Barisone, and the nineteenth-century editor had a hard time making sense of the letter.[51] The letter from Barisone had been addressed to one Bernardo, pastor of the Venetian church of San Giovanni in Oleo, who had "spontaneously brought it to the Tribunal of the Heads of the Council of Ten," then passed it on to the archives of the Savi to be inserted into a volume dedicated to the documentation of Jesuit plots. The intercepted or leaked letter reveals that Barisone had been informed of the trap, which he calls a "diabolical invention" designed to sow discord on behalf of Satan. Barisone sought to remedy the effects his missives were creating in Venice by staking out the ethical high ground in the process of entrapment: "Consider which is [more] shameful for God and for men, deceiving or being deceived?"[52] The fact of the false identity of Cecilia Contarini had been revealed to him "by various sources," and also something of the supposed authors, but not, it seems, Sagredo's name. He asks for that name, supposedly so that he might pray for the sinner's soul. The letter entered the Venetian archives on 2 August, the same day Sagredo sailed away to Syria.

In fact, a draft of a letter to Barisone from the general of the Society in Rome, dated 5 July 1608, shows that Barisone was already aware of the "artifice" [artificio] at least two weeks previously. The general rejects the plan, presumably proposed by Barisone, to elicit the help of an unnamed cardinal in dealing with the matter.[53]

Sagredo finally announced the hoax to Barisone in a letter from Cecilia Contarini's messenger announcing her death, which concludes with the piece's preposterous punch line:

> This letter announces to Your Reverence the death of the most devoted Cecilia Contarini, who has left your Society 5,000 ducats, to be paid in full as soon as you've received news of her safe arrival in heaven, as agreed in the contract that she drew up with Rocco Berlinzone your legitimate representative. Your Reverence will please, with the first mail to arrive from heaven with the said notice, forward the authentic documentation to render due satisfaction.[54]

A brief description of the entire project was provided in the Marciana copy. It is unclear whether this section was originally written by Sagredo, as he is referred to in the third person (as both an "author" and an "inventor," though this may merely document another framing fiction), or by another party as a potential introduction to another published copy. Were the latter to have been the case, Sarpi is the the most obvious candidate as author, as we know that he was aware of the entire hoax, and that in his letter to De l'Isle Groslot describing the scandal he referred to the Jesuits as "foxes." The entire section employs Sarpian vocabulary, but it remains to be seen whether this is unique to him.[55] Sagredo also uses the "fox" image himself, and it seems to be somewhat of a convention: Aesop, Cicero, and Machiavelli all provide precedents for the image of the cunning, untrustworthy fox. A sharp contrast with the forceful lion, perhaps specifically in this context the evangelist Venetian lion of St. Mark, is implied.

A reference to the events taking place "last May" means that the letter was written soon after the originals were produced and archived. There is further evidence towards Sarpi's potential involvement in the satire: in an interview with Christopher von Dohna, advisor to Christian I, prince of

Anhalt-Bernburg, who in 1608 was forming the Protestant Union, Sarpi seems to refer to the recent hoax when he says, "I've got material on the Jesuits, the likes of which have never been seen or published. The stuff on Barisonio [sic], which I helped with, is nothing compared to what's in my hands."[56] This Jesuit material would seem to survive as documents now archived in two volumes of the Archivio di Stato, Venice, in the section "Consultori de Jure." Volumes 453 and 454 contain a wealth of material, collected by Sarpi from 1605 to 1616.

Volume 541 of the same archival section is a collection of documents providing evidence against the Jesuits and records of the Venetians' motives and justifications in expelling them. Its title gives an idea of the polemical intent of this collection: "Writings and Notices from various sources concerning the insidious machinations and evil deeds of the Jesuit Fathers against this Most Serene republic."[57] Sarpi may have intended to print the dossier: in September 1607 Pope Paul V wrote to the Venetian ambassador saying that Sarpi was planning to publish a volume on the Jesuits with Roberto Meietti.[58]

Volume 541 was archived with a particular purpose in mind, as is revealed by its subtitle, dated 14 June 1606: "These must always be read in the College and in the Senate whenever the return of the Jesuit Fathers to Venice or other territories of the State is discussed or proposed."[59] The collection was left open, however, with new documents added right up until the readmission of the Jesuits in 1657. Sagredo's hoax scenario was all too plausible: in March 1609 Wotton described "a late accident which hath yielded us here much sport": a rich Venetian widow had left the Jesuits ten thousand crowns, plus interest, to be paid upon their return to Venice. Commentators observed that as the sum grew in value, her trustees would become less and less likely to want to or be able to part with it, so that the will would actually produce the opposite effect of that intended: the family would always veto Jesuit readmission.[60]

Genuine letters such as these were the template for Sagredo's parody; the originals were both circulated and archived in Venice, generating Sagredo's controlled experiment as they passed through Sarpi's hands.

> Their universal fame is nowadays spread even to the furthest unknown parts of the world. The Jesuits' own designs and interests are always besieging the life and the estate of princes, and the goods and honor of citizens. This has made people who are curiously inclined towards them and not well acquainted with the common opinion about them start to wonder and wish to understand the truth more explicitly as far as they are able. The means were imagined and put into effect, and even though the silver fox being tracked [Antonio Possevino] wasn't yet captured, very industriously another quite malicious one [Anto-

nio Barisone] was found, and most capably went for the bait. Just when he thought he would have to make extra room in his own traps, he was ensnared and became the ridiculous prey of this gentleman.[61]

Jesuit gullibility is described in a favourite Sarpian metaphor: sudden blindness from an excessively bright light. "Poor Father Antonio Barisone, who found himself at that time rector of the College, by the unexpected splendor of one he thought to be the clearest light of highest hopes was dazzled, or rather blinded, and allowed himself to be led, blind and dulled, and fell most carelessly. . . ."

Sagredo concluded the affair and its publication with a vitriolic summary and commentary of the exchange that displays the full depth of his personal hatred for the order and their perceived effect on Venetian political, spiritual, and cultural life. Central to Sagredo's vision of Jesuit destruction was the image of the masked Jesuit; this very practice, as Menini's ode makes clear, had to be turned against its users: "Who will [. . .] teach with equal art / How to defeat deceitful arts?" [*pari et arte docebit / Artes dolosas vincere?*]. Lest the Jesuits assume that they have merely been outsmarted in a show of deceitful cunning, Sagredo carefully points out his deeper ambitions:

> You have not yet seen where my thought is heading, but it surely rests on a foundation of truth that, with the material you supply, is able to raise to such a towering height that even from far off, the monstrous construction of your ugly and heavy operations is visible. You wear yourselves out trying to hide these plans, even if you dress as cunning old foxes, but this time you won't get away with it, and you know that you have left your tail caught in the door, and in a manner in which many will know if you try to come fetch it. A plot has been woven and cast over you that will last forever concerning the witnessing of your iniquity. Happy is he who knows you, for he will stay away from the traps of your deceptions, but pitiful and misled is he that trusts you, for he stumbles along the narrow paths of the imposture by which you live and ends up the poor victim of your desire and greed, the chasm of your vast and never-ending desire for the labors and sweat of the exhausted makes others grieve and weep over these afflictions and disasters; while you frolic away and celebrate their ignorance and simplicity (ibid., f. 70v–71r).

The full extent of the trap is gloatingly revealed, and this declaration too enters the narrative it describes. The revelation was presumably not sent just to Barisone, who is named as one of the actors in the affair, but to a superior, perhaps even the general in Rome (though I have found no trace of the original in the Jesuit archives).

Talk to that cowardly Jesuit whose letters of advice and drafts of the will have all been conserved by a learned person who wanted to know firsthand [*per esperienza*] what your reputation heralded. The letters were procured and gathered with great industry and wisdom . . . Running headlong into the abyss of their own errors, the Reverend Jesuit Fathers have been hatefully chased away from one state after another as scandalous upstarts, seducers, machinators, impostors, and perverse advisors to poor, simple creatures . . . (ibid., f. 71r).

Sagredo does not merely describe the trap; he guides the reader again through its hidden mechanisms and makes explicit the moral consequences which, he claims, it necessarily produces. The rhetorical opportunities the harangue offers are mercilessly seized; at one point Sagredo seems to be writing directly to God. "Did you send the Son to earth to do this?" he asks. "They profess the life of Christ for this and usurp him and confuse their own name with that of Your Son" (ibid., f. 72rv).

Sagredo's favorite lesson to draw from any anti-Jesuit narrative was the evil nature of their political aspirations towards universal monarchy: "What worse outrage could they commit against Your incomparable goodness (which you have seen, being down here too) than that of the whole world one under crown, with a most bitter and keen sword." The dangers of a perceived Jesuit attempt to take over the world is a central trope in Sagredo's political writings, and was seen as the Society's most direct threat to Venetian autonomy and identity. The threat of Jesuit omnipresence was seen through a lens which placed the Society low down on a Venetian global hierarchy of stereotypes. Sarpi liked to compare the Jesuits to Muslims, the assumption being that both were overinvested in ritual to the point of idolatry. When Sagredo went to Syria he claimed to see the same thing, and in his harangue on the hoax he made a similar comparison: "The Arab nourished in the desert or the Scythian raised among beasts would have withdrawn when he heard someone talking as you did to that witless woman." Unlike Jesuits, even tigers and lions, let alone a "barbarian," would not want "to suckle the blood from her veins"; the only creature Sagredo regards as a worthy comparison are the Harpies, "who eat up flesh, bones, and marrows, but only because of prophecy" (ibid., f.74r). The result of Jesuit practice has led to a reversal of values: "Because of your works, virtue is turned into iniquity, faith into idolatry, sanctimony into hypocrisy and charity into violence and rapacity . . ." (ibid., f. 74v).

Sagredo declared himself deeply ambivalent about the fictional world he had invented as the setting for his trap. He refers to one character, a priest from Chioggia, as "introduced with great pleasure into the present comedy" (ibid., f. 78v), but then qualifies the classification of the genre of the piece: "A pleasant comedy, for sure, for those in the know about your dealings, and aren't hurt by them, a refined taste. But it is a funereal tragedy for everyone else whose eyes were veiled by your hypocrisy as they were led to the precipice and unhappily gave their blood and soul" (ibid., f. 79r). Sagredo's own inventions are justified as a remedy to the enormous lies perpetrated by the Jesuits. These fictions extended beyond local politics to the correct representation of papal authority and its relationship to God: the Jesuits, Sagredo claimed, encouraged "impudent, incensed adulators to call him [the pope] God on Earth, Ruler of Heaven and Earth . . . " (ibid., f. 79v).

The Jesuits' peculiar vow of obedience to the pope had created the ridiculous notion of papal infallibility, which had recently been disproved in the political arena by the farce of the Venetian Interdict. Such idolatry in fact led precisely to the opposite effect, for the overinflated estimations of papal power, which the Jesuits esteemed as able "in a second to destroy kingdoms, make and unmake the world, with just a single word," must logically even outdo the power it supposedly represents: "All that's needed is for them to want to subsume the world and he [the pope] is able to move the heavens, the planets and God himself" (ibid., f. 80r). Sagredo's Jesuits have become theological and political parodies of Archimedes, claiming to possess the point from which all existence might be manipulated.

This supposed thaumaturgy is then immediately ridiculed. "I am amazed," he says, "that you have not yet said that he [the pope] can dry the seas, enlarge mountains, raise riverbeds, block out the sun, lighten darkness. Poor us, unhappy us, but you, plagued by misfortune, are more poor and unhappy, when you hear worthily the merit of your works" (ibid., f. 80r). The Interdict experience proved to the Venetians that the papacy and the Jesuits trusted their own propaganda too much: "Oh how evil, stupid, and unfeeling to believe with your currency of invented histories that you might terrify wise and intelligent men. Liars, don't you see how in the end your machinations would fall ruinously on your own heads with perpetual disgrace" (ibid., f. 81r).

At the heart of this miscalculation was a vain supposition that the order had the right "to speak as secretaries of God" (ibid., f. 81v). This was not merely presumptuous; it presupposed a "divine arcana" jealously guarded from the masses by a secretive God who, however, was prepared to allow access to the Jesuits. This idea of privileged access to the divine court, Sagredo argued, gave the Jesuits the impression that they were merely passing on divine judgement as messengers rather than actively meddling in geopolitics and manipulating history. Sagredo illustrates the Jesuits' misunderstanding

of politics by offering them examples from their own attempts to guide recent history, comparing the failures of the Jesuit-guided Sebastian I of Portugal to the successes of the recently deceased Protestant Elizabeth I of England.
(ibid., ff. 82v–83r.).[62]

Failed machinations cause failed hermeneutics, and these have disastrous
consequences:

> You joke, you twist, you scheme, you envelop with your writings, taking two
> words from the middle, four from the end and three at the start, you strike
> with insidious temerity all at once, so that you are not just a monster, but a
> fetid, misshapen defect that throws itself into disorder and leaves itself incom-
> plete so that you are left unable to produce further offspring: nowadays you
> like to act in one direction, then another; for your interests and tastes you vary,
> and when these change, so must your comments change the sacred scriptures,
> which are always one, always sacred, and whose judgment and admiration
> must rest always the same, too. You wish that these would adapt themselves
> to the unstable winds of your affections, but this won't do, so your sects force
> themselves to jest and mock, twist, envelop, and sow the field afresh.[63]

The harangue is necessarily high-pitched, but its own strategies of attack
are varied and unpredictable enough to make it a challenging and uncom-
fortable read. Perhaps the only pillar of virtue in the piece is the deceased
senator Antonio Querini, whose *Aviso* (1606) had drawn harsh criticism from
Rome.[64] As Querini was a friend to Sarpi, Sagredo, and Galileo, it seems the
Jesuit attacks on his character after his death in 1607 might well have provided the local context for conceiving the hoax.

The immediate effect of the exchange is unknown: Sarpi's letters to Groslot de l'Isle and Foscarini, Menino's poem, Sagredo's references in his extant
letters to Galileo, Barisone's letter to the pastor at San Giovanni in Oleo, and
the Marciana copy are all that we have to go on. Sarpi, Donà, and Contarini
certainly continued to resist Jesuit overtures campaigning for their speedy
readmission to Venetian territories. On 18 August 1608 it became illegal in
Venice even to correspond with Jesuits.[65]

The importance of the exchange is twofold: this may not be *A Modest
Proposal*, but it is satirical in intent and effect. Historians of science have
tended to miss or remove their actors' sense of humor. This may be because
the central figures of conventional narratives of the scientific revolution,
such as Copernicus, Boyle, and Newton, did not seem even to their contemporaries to possess any. But many others, such as Kepler, Galileo, Pascal,
Kircher, and Hooke, did—and we would do well to integrate humor and

its histories among the resources we use for writing intellectual history.[66] More concretely, for Sagredo the exchange with "Berlinzone" did not end in the summer of 1608: this was merely the start of a campaign whose aim was not necessarily to insult the Jesuits as much as to subvert their information-gathering networks for his own ends.

Interconnections

Little trace has remained of Sagredo's venture in Aleppo, but it is still possible to reconstruct its intended form. His three-year relocation from Venice to Syria was not a departure from politics or natural philosophy. His few surviving letters from that period make it clear that he considered his new geographical location an opportunity for producing new knowledge. In reality, there is little evidence that he actually managed to observe or experiment in Aleppo. What we find instead is a redeployment, already formulated in Venice, of a new anti-Jesuit correspondence network unwittingly deployed by Jesuits against themselves.

In both Sarpi and Sagredo's violent rhetoric, the Jesuits are repeatedly accused of machinations and manipulations. They are frequently referred to as "instruments," passive actors moved by unseen forces, mere reifications of deeper, hidden, diabolical plans. The machinator also transforms politics itself: discursive humanist interventions or natural husbandry give way to dichotomies of surface and subterfuge; contemporary politics cannot be read by observing mere events, their arcana must be penetrated, unveiled, disclosed. This notion of history and politics also applies to nature itself. While Sarpi exposes, anatomizes, and eviscerates, Sagredo adopts a different approach, intervening, infiltrating, and co-opting.

Perhaps the best way to experience Sagredo's activities is to reproduce in full two of the three personal letters to survive from his time in Aleppo.[1] Both were written on 30 April 1609: one, addressed to Sarpi and archived in a bundle of correspondence, largely unpublished, was discovered by Gaetano Cozzi but has not subsequently been studied by Galileo scholars.[2] Written alongside a letter to Galileo which was sent at least in duplicate, Sagredo's letter to Sarpi gives us a glimpse of his machinations in world systems from his new base in Aleppo:

My Most Reverend Signor,

I have delayed writing to you in the hope that one day I might have some leisure to follow better my thoughts, but as it's already been seven months that I have been leaving off this hope in vain, with more and more business every day, I wanted at least to inform you of the reason of my silence, especially as the same reason has prevented me from discussing with you the things I intended to. Here the material for philosophy is infinite and I fall apart in the desire to have you here with me along with Signor Galileo, because I know that were we to have all our time free for speculation, we would pass our lives with incomparable pleasure.

I have assembled a project in far-off countries, and if I receive responses to the questions and queries I have made to various people, spread over the Indias and other places in Asia, I do not doubt that I shall have material to form volumes of extremely curious accounts. But this business takes up a huge amount of time due to the mailings I have to put together for all these places, not without relevance to my wallet: with the diligence of writing I reply, triple, and quadruple my requests, taking care to invite my correspondents to the same diligence, so it is necessary that donations and offerings flow everywhere to increase, for my part, my reputation, and for the others, their debt. I cannot believe that with more than twenty-five subjects that I've set working in my research, four or six at least won't feel the debt of duty; and who knows if the Jesuit Fathers won't be the only ones to come up with something to match my desires? If God wills that I return home, how much fun will all our friends have reading the collection of letters sent and received from these Fathers? I'm hoping that the trust we'll see in this matter between them and me might go beyond that which we saw between Berlinzone and Colomba; but what makes me happiest is to imagine to myself that even after my return I will continue the project, receiving continual accounts from the Indias, about which I have heard from several people three details concerning the Jesuits that are very important. First, that in the Indias the Portuguese Governors do not use ciphers apart from in times of war, and ban others from using them, conceding them only to the Jesuits, who say that it's necessary for the greater service of God, and they take full advantage of this, from which you can understand how much better their Apostolate might be compared to that of the first twelve Apostles. Furthermore, after that of the King, the greatest riches in the Indias belong to the Company of Jesus, and I have been told that they possess not just enormous lands and rents everywhere, but that in particular they rule almost all the island of Salsette [the site of modern-day Mumbai] towards Khambhat [in Gujarat], which they say is about twenty leagues round, and also that the Jesuits command several fortresses, where they levy taxes and every other duty proper to princes, placing soldiers, Governors and Judges according to the normal administration of politics. Towards Cape Comorin in the kingdom of Calicut they possess the fortress of Tuttocurè. Between Chaul and Vasai they have the fortress of Bandra, and I have even heard that they have Bombay, but have not been able to verify this. Concerning the progress they have made in Japan and other gentile places, where the Portuguese keep fortresses, and there is no prelate, they reap great benefits and fame, for they are the heads of the Christians, and exercise spiritual authority in these countries, and the Portuguese themselves led there for business are more obedient to them than the others, bettering their condition with their obedience. I have also been told (and it seems quite credible) that the Christians submit to their judgment on their disputes, thinking that it is a sacrilege that one Christian take another to the justice of infidels. In India there are other religious orders of St. Francis, St. Augustine, and St. Dominic, but they do not have the credit of the Jesuits, because beyond their spiritual ministry they have no jurisdiction, and do not deal in other business, and their needs are taken care of by the Portuguese with the assignment of soldier's pay, or something similarly limited, and in addition the Jesuits aspire to inheritances and similar things, and managing the goods of the confraternity of charity, they proceed mainly by insinuating themselves into everyone's thanks, and display exemplary sanctity, and if one of them swerves from this good name that they procure, they throw him two, three, or four thousand miles away, and as the fame of the error diminishes, they spread the story that the culprit has been severely punished, which keeps the whole company in incomparable veneration. Please excuse me if I cannot write more, because for tomorrow evening I will have had to send more than fifty letters. I affectionately kiss your hand and pray God give you every happiness, as I do too to Father Fulgentio,[3] and the whole company.

In Aleppo, the last day of April 1609.

Your Most Reverend etc.

Most desirous to serve,

Sagredo.

Compare this with Sagredo's first letter to Galileo from Aleppo:

Most Illustrious and Excellent Signor,

I talk, I discourse, and am with all my soul always with you, nor after my arrival here have I been able, or known how to write to you. Not for a lack of material (because there are so many novelties here, and opportunities to philosophize, that I can't take two steps without wishing you were here with me to hear your opinion), but because, from the other direction, infinite business and disruptions (and these mainly tiresome and bothersome) distract and worry my soul so that I remain unable to write to you as I would like. So, as

I don't ever see the moment coming when I might write to you with leisure, I wanted at least with this to rid you of the wonder you may have given my silence.

Here I have awoken such a burning desire to know infinite things that I curse my ignorance and time lost in leisure, which I should and could spend studying, a thousand times an hour. If you could see me in my study going to open and leaf through books, I know that you would laugh, observing how while I, drawn by curiosity, open one of them, I really want to study another; and as if I were afraid that it would run away, I am compelled by a strong desire to take it, and after that one, another and another, until I am loaded up like an ass. At last, giving myself over to reading one, the thoughts and business continually running around my head make my tongue and eyes tired of reading, without my intellect being able to understand anything; and if by some chance I pick something up, my memory, distracted by troubles and needs, doesn't know how to retain it. So my studies consist only of a burning desire, separate from the intellect and memory, which, ruled like a tyrant by annoying work, is left completely incapable of granting them audience. Nevertheless I console myself with the hope of being with you in Padua for a couple of months to philosophize and have fun, but at the same time the thought that at least three years must pass before this longed-for effect worries me inordinately, and that the dangers of such a long voyage prevent me from being sure of my return. And this last barrier seems to wear down my hope more than the mere length of time, as a hundred years, assigned as the outer limit of human life, seems to me a brief space, I know that three will pass all too soon, and that with them will pass all too acutely a good part of this life's vigor. Make do with this while you console me with your laughing letters, and know that, blinded by the pleasure I enjoy reading them, I trick myself into believing you present. Alas, that work prevents me from engaging longer with you, to whom in the end and without end I recommend myself, praying God send you every happiness and joy.
In Aleppo, the last day of April 1609,
All All [sic] yours,
Sagredo.[4]

To Sarpi, Sagredo presents himself as a nascent intelligencer, constructing a durable and global information web and offering up its first catches. Syria here is not the end of the Mediterranean; it is the start of Asia, opening up vivid cartographic vistas of Jesuit conspiracies infiltrating the Portuguese fortresses of west India, extending their noxious influence across to Japan, poisoning other missionary orders and ultimately destroying Venice's access to the rest of the world. Sagredo's solution, hinted at in other letters, is to turn the mechanisms by which the Society lived against itself, continuing and expanding the Berlinzone-Colomba exchange with higher stakes on a global scale. His insight, perhaps inspired by his reading of Gilbert, who theoretically turned the globe into an instrument, was that some scientific problems were geopolitical problems. The solution he proposed has a strange logic and a stranger history.

Reliable global data required reliable global observers; the only globally distributed organization in existence was the Jesuits; and therefore the Jesuits must provide scientific data. This is indeed part of what happened over the long seventeenth century: the Society of Jesus, rather than any state-based empire, attempted to address the challenges of coordinating and standardizing global observation, especially astronomical observation.[5] Historians have, however, largely misunderstood the nature and chronology of such a development, projecting back onto the late sixteenth-century Jesuit *Annual Letters* a centralized and disciplining scientific force which did not yet exist. In reality, however, the emergence of global observational networks came about both internally, from the missionary system, and externally, from the pressures of patrons to appropriate the Order to their own ends. Sagredo's satirical efforts to co-opt the Society in Asia predate the French érudit Nicolas Claude Fabri de Peiresc's less humorous project, which in turn inspired Jesuits such as Kircher to use the network from within.[6] The origin of the Jesuits' global scientific network was satirical conspiracy theory.[7]

This is all the more strange, given that historians have tended to follow what we may term Sagredo's "black legend" assumptions in viewing the Jesuits as the only religious order with the potential to produce global science. To other contemporary observers this was far from true: Carmelites, Capuchins, Augustinians, and Franciscans were all widely deployed, and, while they may have lacked the centralized scientific infrastructure of the Jesuit's Collegio Romano, this organ's importance for missionary science may have been overstressed in an idealizing historiography.[8]

The subject of this new experiment was one that bound Sarpi, Galileo, and Sagredo together: magnetism. Sagredo's tests in 1602 of Galileo's "declinatorio" had provided precisely what he also claimed to find in Aleppo in 1609: "material to philosophize."[9] Sagredo finally got around to fulfilling his role as data-gathering scientific diplomat in October 1609, writing to Galileo from Aleppo:

> I have carried out the observations of the magnet, which most certainly declines here by seven and a half degrees towards the North-West, so that from Venice to here the difference would be fifteen: off you go now to investigate the reason. I sent a good needle to the Jesuit Fathers of Goa, praying them to make a precise observation with it; and I hope to have with them the same

correspondence Colomba had with Berlinzone. Actually, I get more letters
from them than from you, from whom I've had in a year just one, plus one
from the king of Persia, and I want to see from whom I get the second.[10]

The only other direct statement we have regarding the content of the correspondence comes from 1617, when Sagredo wrote to Galileo and the subject
turned to fruit. "I wrote," he said after a discussion of Syrian dates, "when I
was in those parts, to the brothers of M. Rocco Berlinzone in the Indias, for
them to send me flower seeds or other plants that aren't in Italy; but from
them I received nothing but guff and promises."[11]

So no botanical samples traveled through his web, but does this mean
the web itself failed? In his first letter to his friend after his return to Venice
from Syria—a letter famous for its stirring warning of the perils of attempting to think freely in Jesuit-controlled Tuscany—Sagredo offered a cautious,
ciphered invitation to Galileo: "In India I was in close correspondence with
the brothers of M. Rocco, and I have another small register to add to that of
Madam Anzola Colomba."[12]

This was clearly meant as a nostalgic lure of common anti-Jesuit satire,
but we have no record of Galileo's response, nor of the nature of the hoax.
There is, however, one faint echo of this Indian exchange in the Venetian
archives. In the same volume of mixed Sarpian writings already mentioned,
"ASV, Consultori in Iure, 453," where the Sagredo letter to Sarpi is archived,
there used to be a small collection of twelve folios whose title, recorded in
the early seventeenth century, is "Lettere diverse scritte da Giesuiti e fratti di
S. Francesco dalle Indie Orientali" (Diverse letters written by Jesuits and
Brothers of St. Francis from the East Indies). While Sarpi had direct correspondence with a large number of intellectuals across Europe, he tended
to get his global information secondhand, and so it is unlikely that these letters were addressed to him. Sarpi mentioned the provenance and content of
this material in a letter to Groslot de l'Isle in July 1611, just a few weeks after
Sagredo's return from Aleppo via Marseille and Milan: "With the occasion of
one coming from Syria I have understood a great deal concerning the Jesuits'
progress in the Indias, where they have reduced themselves to open domi-
nation: a clear indication of their intention for the same in Europe, if they
can."[13] It seems likely that this represents the last known trace of Sagredo's
Indian exchange, filed away amid responses to Suarez, political tracts on mo-
narchical supremacy, and *avvisi*.[14]

Whatever the eventual results of the exchange, Sagredo's accounts to his
two friends are striking for the different directions they take from similar
openings. For Sarpi, the weight of business leads to a new world of insights:

philosophy is sacrificed for information gathering and control. For Galileo, the same weight leads instead only into a disordered study and thence into a distracted mind, incapable of functioning without the corrective guidance of Galileo's Paduan discipline. The mass of correspondence generated for Sarpi, with fifty letters due the next day (it is unclear whether this figure is a typical Sagredo exaggeration, given the general tone of the letter) is sharply contrasted with the failed consumption of piles of books. Nature itself, in its Syrian manifestation, is, for Sagredo in this account, both "material to philosophize" and incapable of being studied without Galileo. In the letter to Sarpi, Sarpi's presence is also deemed necessary to produce philosophy, but the full vain fantasy of socioepistemological plenitude is to imagine a constant reenactment of the Barozzi banquet, with the Sarpian coterie again permanently enthralled by Sagredo's daring subterfuges. With Galileo, a master-student dialectic is posited as the ideal mode of interrogating nature; with Sarpi, both the production and reception of knowledge are collectively distributed. With Galileo, the letter performs the humanist task of making the absent present; with Sarpi, it probes and infiltrates, working as an instrument.

Sagredo's submission of his measurement of magnetic declination to Galileo later in the year shows in fact that he was capable of "doing philosophy" in Syria: the opposition between his work and the freedom of the spirit (*animo libero*) he longs for starts to break down. We see already in the letter to Sarpi that such a stark division is rhetorical. The same techniques of bureaucratic paper management are deployed in consulship as in philosophy. In fact, philosophy, even subterfuging philosophy, depends on the writer's diplomatic status in order to work. The quadruplicate letters, the system of gifts, the careful calculation of obligation, the distribution of work to the letter—all these are the tools of the diplomat.[15]

There is a further possibility that Sagredo himself received a telescope as a gift from Galileo while stationed in Aleppo, and that he conducted the first telescopic astronomical observations outside Europe at a remarkably early date. The single piece of evidence upon which the vague claim might be based is extremely difficult to interpret, with curiously garbled grammar: writing in 1612 after his return from Aleppo, Sagredo offered up this account:

> I have not observed the Medici planets; actually being in Syria I observed the Medici stars with the first Instrument that I had, indeed before I had it it was [or "you were"] waiting in great anticipation to observe the same constellations, that you in fact have observed: whence then reading the *Sidereus nuncius* I was left in some wonder at having come across so precisely the same part of the Heavens. If your observations of the aforementioned planets will be sent to me from you it will be the cause that I will observe them.[16]

Sagredo does seem to refer to a telescope, with which he made astronomical observations while failing to notice the Jovian satellites. As he does not thank Galileo for this instrument, it is likely that it was a low-powered spyglass obtained through mercantile or military networks in the eastern Mediterranean, which would have been capable of making more stars visible, but not the satellites.

Spying

It is perhaps instructive to glimpse Sagredo's other activities in Aleppo through the remaining archival traces in order better to understand the position this epistolary philosophy occupied within his world. In 1609 and 1610, Sagredo intercepted and copied a series of documents in Aleppo which he sent back to Venice. Such espionage, as we shall see, earned him a sharp reprimand from the Council of Ten, but the delays in transporting documents across the eastern Mediterranean meant that the practice of diplomatic spying continued.

The files have long been invisible, as they are archived under the wrong date: 1709–10 instead of 1609–10. Paolo Preto first rectified the mistake in his extensive study of Venetian espionage.[17] They are also misordered within the volume, and each self-contained bundle contains several copies, transcriptions, and translations of the same set of documents. Their eventual shelfmark, as part of the archive devoted to the State Inquisitors, is also part of their story, as the documents were passed from one body of government to another as they accrued secrecy.

The documents are worth describing, not merely to reconstruct Sagredo's interventions in global geopolitics but to understand his attitudes towards documents themselves. Sagredo's cover letters describe in detail the practices of textual theft, transcription, translation, and decipherment that generated the volume. Such practices were transferred to the realm of natural philosophy, where they underpinned both the hermeneutics of nature and the management of documentary culture that allowed, and sometimes impeded, natural philosophical knowledge to be produced and replicated.

The volume consists of two main collections of intercepted documents, plus copies of the correspondence between Aleppo and Venice about them. The first collection, which Sagredo sent to the Council of Ten on 2 September 1609, contained letters from the Carmelite mission in Isfahan as well as other more sensitive correspondence. The letters were being carried by Shah Abbas I's envoy to Venice, the Armenian merchant Xwāje Ṣafar, from Julfa, and were addressed to various European leaders. Some were written in Persian, which Sagredo could not read, so he merely copied them as best he could without having them translated.[18]

The first letter, from the Carmelite friar Giovanni Taddeo to the general of the Order in Rome, was written in Isfahan on 30 January 1609 but was sent, with various postscripts, on 2 May. The letter demonstrates just how intertwined missionary issues were with diplomacy and trade: the Carmelite mission had been sent to Persia by Pope Paul V as a response to an Augustian mission sent by the Portuguese on the initiative of the viceroy of India. The tensions between the two missions, and the conflicting interests of their two backers, rendered both missions compromised and weak. The quixotic schemes of the papacy to attack the Ottomans with a European league coordinated with a separate attack by Abbas had left Persia feeling betrayed, isolated, and overstretched.[19] Abbas exploited the tensions between the missions to punish the papacy and Spain by proxy. Taddeo relates how after being allocated one of the best houses in Isfahan by Abbas, complete with running water, he was suddenly turned out by a squadron of 150 soldiers who transformed the Carmelite church into their stables, and was told to share the quarters with the Augustinians, who then forced them to move to an inn. The "instability of this king and kingdom" made running the mission nearly impossible. Nevertheless, one of the Augustinians received permission to found a monastery in Portuguese Hormuz. Abbas had taken Bahrain in 1602, and clearly wanted to remove the Portuguese from Hormuz. It was eventually taken, with English help, in 1622. The island was deeply contested, and Shah Abbas's granting of permission to the Carmelites to set up a monastery there may have been less munificent than it first seemed: reports quickly came back of Portuguese resistance to the Italian and papal mission.[20]

The competition between the Carmelites and the Augustinians created different approaches to the running of their missions: the Carmelites set up a school where they taught fifty Portuguese children to read, write, and sing; the Augustinians took to preaching in the streets. Local Christian populations were also treated differently by the two groups: the Augustinians barred Armenian Christians from their church, calling them schismatics, which the Carmelites said was untrue (ibid., f. 33r).

The missions were not mere victims in the game of international diplomacy, however; they were also actors. Taddeo claimed that Shah Abbas "loves merchants, not friars," and he was instructed to carry letters of recommendation to the pope, the doge of Venice, and the grand duke of Tuscany for Xwāje Ṣafar (the bearer of these letters to Aleppo), who had been entrusted to retrieve goods belonging to Abbas from Venice, and to engage in other trade. Copies of Taddeo's letter to the pope and the grand duke explaining Ṣafar's

mission are included, as well as a letter to the papal nuncio in Venice informing him of Shah Abbas's annoyance at the peace treaty between the Holy Roman Emperor and the Ottomans (ibid., f. 34v–35r, 35r–35v, and 35v–36r). The
bundle concludes with Sagredo's transcriptions, untranslated but with the
identities of their recipients noted, of the Persian documents: Ṣafar's commission and letters from Shah Abbas to the pope, the king of Spain, the doge,
the grand duke of Tuscany, a priest, and Anthony and Robert Shirley.[21]

Another document was, as Sagredo realized, more dangerous to inter-
cept. Sent from the Quirimbas Islands off the east African coast on 20 June
1609, it was a letter from the viceroy of India to the Spanish king. The letter
was heavily ciphered and originally in Portuguese, a language Sagredo did
not understand; yet he managed to both decipher and translate it. It contained important and desperate military information on the limited Spanish
naval and armed forces in India, revealing that there were only three Spanish
galleons and "few people, of no consideration when hoping for service, both
nobles and soldiers."[22] The Portuguese residents lacked money, having spent
it all on the galleons and taxes; the Dutch had been spotted in Mozambique;
monsoons made military or mercantile expeditions impossible.

The second group of documents was sent in February 1610. Sagredo wrote
in cipher to the Council of Ten in Venice with an urgent bundle of intercepted documents. A Portuguese subject had arrived in Aleppo from India,
robbed of his goods but bearing letters to the king of Spain concerning Persia. Sagredo hosted him, opened the letters, copied them, and resealed them.
A few days later a courier arrived, also on his way to Spain but carrying some
early replies for Sagredo from his letters to India and Baghdad. The courier,
impressed by Sagredo's reputation for generosity towards Portuguese subjects, gave an account of news and showed Sagredo the two envelopes he
carried: one for Spain, the other for Venice. Sagredo pretended to have no
interest in these, but then stole them, copied them, and resealed them. The
letter to Spain was ciphered; the other, addressed to the Spanish ambassador
in Venice, was not.

The first letter was from the Augustinian friar Antonio de Gouvea to
Philip III, king of Spain, dated 15 April 1609. De Gouvea was a key player
in European-Persian relations: he was first sent to Persia in 1602 on the authority of the Portuguese viceroy of Goa with the unenviable brief of convincing Shah Abbas that the Europeans were serious about coordinating an
anti-Ottoman war on two fronts. De Gouvea returned to Goa in 1603 and
published an account of Persia, based both on his own experiences and other
Augustinian reports. In 1608 he returned to Persia to set up a permanent mission, bringing with him a letter from the king of Spain. Abbas then sent de

Gouvea to Spain and Rome; he arrived in Madrid in 1611. The purpose of this journey was both military and diplomatic: Abbas, disappointed by the Europeans' failure to act on their anti-Ottoman pledge, sought a more formal treaty; he also offered to bypass the Ottoman control of European silk supply by opening Hormuz directly to Spanish merchants.[23]

The geopolitical stakes were high, as Sagredo realized. The letter to the king of Spain he intercepted was probably only one of several copies sent via different routes, and Sagredo had no way of knowing whether the information was still relevant, or even true. It recounted in detail the peace treaty between Persia and the Ottomans, and a graphic description of Abbas's personal grief at having been lured into a Turkish war by treacherous European powers. Most important for Sagredo, however, was the news that Abbas intended to transfer his supply of silk from Venetian to Spanish merchants, and to move the market to the crucial port of Hormuz. Threatening these developments was the sighting of a Carmelite friar, sent by the pope, in Hormuz.[24] The Carmelite mission was not only in competition to that of the Augustinians; it represented Roman rather than Spanish interests.

Sagredo also sent two long letters. The first, addressed to the doge, is now in the archival section of official correspondence, where it is placed with a series of documents concerning Ṣafar. Dated 2 September 1609, it was brought by Ṣafar, who was presumably also asked, somewhat audaciously, to carry Sagredo's illicit copies of his own correspondence. In it, Sagredo advises the doge to treat Ṣafar well in order to impress Shah Abbas and to further Venetian interests in Persia. Sagredo adds, in language echoing his discussion with Sarpi of the beneficial effects of long-distance gift giving, that he has lent Ṣafar money and treated him generously, creating an excellent impression of Sagredo in Persia.[25]

The second letter, dated 28 March 1610 and addressed to the Council of Ten, is longer and amounts to an justification of Sagredo's multiple interceptions.[26] This and other cover letters help us understand what it was that Sagredo thought he was doing in stealing private and secret documents; the Council of Ten's replies show Venice's views of his activities.

The most sensitive document Sagredo stole was the letter from the viceroy of India to the king of Spain. Here it is worth recalling the information he had passed on to Sarpi a few months before the interceptions, in April 1609: "In the Indias the Portuguese Governors do not use ciphers apart from in times of war, and ban others from using them, conceding them only to the Jesuits." This information had now been updated: "The opinion that I had that the Spanish, who generally make use of perfect ciphers, was the reason I held off for several days before readying myself to lift the one from the Viceroy of India, which I sent you, being wary of entering into a long and difficult enterprise which would require the mind to be free from other thoughts, but as soon as I started looking at it I found it to be the most simple one can use, and it turned out to be ever so easy to lift it" (ibid., f. 67r). The practicalities of the job, however, do not mask its ethical and political issues, and Sagredo repeatedly expresses his doubts about what he has done. "I do not know how to judge whether in this service I have followed your wishes," he says. Distance itself is a major factor in diplomacy: "Far-off things always appear altered."

But more important is the impossibility of fully understanding another actor's motives: "There is no human operation on earth so well understood, that considered from another direction would not have a thousand oppositions, whence, according to the opinions even of the wisest it might not appear evil and worthy of correction, and all the more so, when there's no one there supplying the arguments from the other side" (ibid.). The role of the consul is itself ambiguous: local merchants assume that he exists for their service, but he is really there to serve and represent the government. Distance renders his adequate communication with the Council impractical, yet his brief is to represent and act in Venice's name. The result is a paralysis: Sagredo fears that either he will appear to be a man lacking resolution or he will risk displeasing the Council. Even a perfectly executed and well-intentioned "operation" might appear imperfect and evil at a distance. In an image recalling maritime lighthouses or Kepler's magnet-like sun, Sagredo asks for the distant "light of your will" to act as the "infallible rule of all my operations" (ibid., ff. 68v).

In the absence of such guidance, Sagredo can steer only on two principles: to obtain and convey full understanding of Persian affairs, and to understand how to use them to engage the Turks. It is not clear whether such policies are of Sagredo's own devising, or which were entrusted to him orally in his brief before his departure. In order to effect them, he regards it as essential to "imprint on the Shah a favorable image of Your Excellencies," without, however, risking leaving traces by which he might be identified (ff. 69v). Representation requires self-effacement. Even an encoded description of his activities within the present letter might compromise his actions, so it is not included; yet such a lack of detail should not be seen as a sign of inactivity.

The doubts, confusion, and paradoxes of Sagredo's letter might perhaps make us, and the Council of Ten, think that he was merely a poor choice for consul. The contrast with the confident instrumentalization of correspondents in the letter to Sarpi is striking. Here we see more complex systems at work, with Sagredo caught between the conflicting interests of the mercantile origins of the consul and the emerging conception of state representation on the one hand, and the conflicting interests and information regimes of globalizing powers on the other. Part of his solution, espionage, is reliant on the Sarpian model of information gathering he had developed in his hoax with Barisone: the more information surreptitiously gleaned, the better. He is at pains to detail his manual skill in espionage: "No scandal can arise from me opening the letters," he assures his superiors, "because I would not have undertaken the task were I not certain to reseal them in such a way that, should the very writer see them, he would not notice anything amiss. I have always been sure before getting my hands on a letter to observe carefully the script, seals, paper, wax, and thread, with the thought that should I need to change something, I could do it safely" (ibid., f. 2v). Even the copying of unknown scripts is put to test: Sagredo experimented with replicating legible versions of innocent texts in "Persian, Turkish, Arabic, and Chaldean" (ibid., f. 30r). This model of pure replication without comprehension underpins Sagredo's entire self-representation: he portrays himself as a passive but privileged amanuensis of international news, both responsible for and innocent of the information he conveys. At the same time, his expressions of doubt concerning the appropriateness of his activities reveal the underlying tensions.

The Council of Ten's controlled reply, sent on 21 June 1610 in response to both packages of intercepted documents, confirmed Sagredo's fears:

> We know, through the great excess of your diligence and vigilance concerning everything that you think might concern our interest or service, that you came to a decision (as you wrote in your letters of 25 February and 28 March, whose copies, and shortly after, originals, recently arrived). If you were to have abstained, we would have considered you incomparably more dear. As the Council of Ten, without descending into further details, we trust that this hint will serve you as enough light of our intentions, which are extremely far from that which went through your hands, and with this we wish to be sure that in the future you conform in and by everything to keep as far away as possible from any material causing displeasure or evil consequences. We appreciate the great value of your ingenuity and we are certain of the excellent ends that moved you in your actions, but we promise much more of this.[27]

Sagredo's expanding dossier was now passed, permanently, to the Inquisitors of State.[28] The Inquisitors then sent another letter to Sagredo on 20 June 1610. Reinforcing the collective judgement of the Council of Ten, the inquisitors also added, in code, practical instructions to Sagredo to cover up the remaining traces of his misguided ventures, stating that the affair had to "remain buried under a perpetual silence." Archivally this meant several things. First, Sagredo's various missives to Venice were removed from their original positions in the files of their recipients and transferred to the Inquisitors' own archives, making access to them and knowledge of them far more difficult. Second, references within existing registers or indexes to the correspondence had to be removed. Third, Sagredo himself had physically to eradicate any trace of his interceptions and communications from his own document collections in Aleppo. No reference or allusion to the exchange was to be permitted in any public letter. The interceptions had never happened; the conversations had never taken place.[29]

What emerges in this correspondence is a constant process of reframing of documents: the meaning of the text, whether deciphered, translated, or merely transcribed, is only part of the meaning of the document. The various repetitions within the volume, with some texts present in four separate copies, thus signals less a bureaucratic multiplication of paperwork than an awareness of the importance of each copy's particular provenance, real as well as pretended, and the texts' change in meaning as they were passed through the hierarchies of government and read alongside other documents. The invisible traces of the stealing, opening, copying, resealing, and replacing of the letters—multiple times, copies as well as originals—rendered the words, Sagredo's prize, the least important part of the materials' meaning. What the letters ultimately signaled was that their own system of conveyance was untrustworthy.

The Venetian Inquisitors ordered Sagredo to destroy any copies of these letters he had retained for his own use. He replied that the only copies he had kept would have been useless to anyone else, but that he had completely destroyed them. Presumably, he had ciphered these using a private code: his letter to the Inquisitors concludes with a comment that he had not been sent the key to the ciphers they and the Council of Ten had employed in their stern letters, with the implicit criticism that their systems were transparently breakable: "We may attribute to good luck that I was able to work out what was written to me by mistake, and it's quite likely that I wouldn't always manage it with other codes."[30]

What we see in this miscellaneous collection of documents is not just a bad diplomat or an amateur spy, but an assemblage of skills, both material and intellectual, which were also deployed in other realms of experience. Interception, transcription, translation, decipherment, replication, interpretation: all these were equally useful, as we shall see, in Sagredo's contribution to the sunspot debates. Other techniques, such as the expansion of the Venetian Interdict hoax using a fictitious correspondent to the global realm of

Jesuit missionary natural philosophers, more literally merge the identities of the diplomat and the natural philosopher.

Another set of documents from Syria further illustrates the documentary mentality informing both diplomatic and natural philosophical practice. Both help us understand not only the daily labors that Sagredo complained robbed him of his chance to philosophize in Aleppo, but also the role such activities played in establishing and dismantling relationships between European mercantile communitites in Syria and the Ottoman authorities.

The first document is written in the form of a journal, covering the period 7 September to 11 October 1609.[31] It recounts Sagredo's investigation into an attempted murder at his residence in Aleppo by Francesco Pavese, his *maestro di casa*, of a Venetian merchant, Nicolo Ciroiso. The proceedings have little intrinsic value, but they do show the investigative mode crucial to residents, consuls, and ambassadors for containing disputes within the jurisdiction of their own nations. Transcripts of interrogations are included, with Sagredo repeatedly exposing inconsistencies and lies in Pavese's account, and insisting on the correct juridical procedure. Pavese escaped a couple of times, and Sagredo had to enlist janissaries to return him; this threatened his attempt to maintain autonomy over the affair. In the end, with Pavese threatening to "turn Turk" if forcibly imprisoned or deported, Sagredo sent him to receive trial in Venice with the Avogadori di Commune.[32]

The possibility of "turning Turk," as well as the host of minor characters appearing in the narrative, ranging from Venetian and Syrian spiciers and barbers to translators, jannissaries, moneychangers, and consuls of other nations, reminds us that the autonomous mercantile "nation" was always a fiction.[33] One of the disturbing crimes committed by Pavese was leaving the doors to Sagredo's residence open at night: the "palazzo" is described as habitually kept "like a fortress, given how many thieves and brigands roam around at night."[34] But the imagery of the fortified compound is belied by the interactions with many non-Venetians during the investigation itself. The consul's power, originally underwritten by the resident merchants, and at this point gradually shifting towards a model of state representation, was always limited: the best Sagredo could hope for in this case was that the suspect faced trial in Venice, and to limit the scandal and loss of confidence and credit with competing nations and the Ottoman regime.

A second dossier is more revealing, both of the conflicts between resident "nations" and the related reframing of an historical event as it became enmeshed in competing contexts and power relations. The central historical event, whose status and meaning were dramatically recast during the compilation of the dossier, seems initially to have been a classic act of political iconoclasm. On Good Friday, 16 April 1609, a Venetian merchant, Alvise Bisutti, walked into the church of the Virgin in Alexandretta and saw, posted on the wall, a printed picture of the French royal family. He ripped the sheet down, screwed it into a ball, kicked it around "as if it were a football," and then, changing sport, batted it out of the window with a stick. This simple act produced a series of reactions, arguments, and processes.[35]

A crowd of French sailors seized Bisutti and imprisoned him aboard their ship. Complex negotiations took place to have him safely released with the guarantee that he would not be set free without a thorough investigation; the French refused to release him, while the Venetians refused to start an investigation with him imprisoned. The helmsman of the Venetian galleon *Emo* (on which Sagredo had arrived in Syria) was kidnapped by the French while fishing, and held for ransom. French scrap paper accounting sheets were apparently found at the scene of an arson attack on the Venetian salt stores. Venetian merchants in Aleppo protested, calling for the affair to be judged by Aleppo's qadi, Mustafa Aburacheb. Sagredo initially refused, asking the English consul, Paul Pindar, to intervene as intermediary with the French vice-consul, but Aburacheb, a friend of Bisutti's brother, directly intervened and had Bisutti freed, with his brother agreeing to be held hostage in his stead. The French requested that they themselves try Bisutti, and gathered witness statements.

At this point, interpretations of the basic events and their supposed causal relations to other events start to multiply. The French tied Bisutti's act to a general anti-French policy, claiming that a year earlier he had been the catalyst of the murder of the previous French vice-consul in Alexandretta, Honoré Gassendi. The actual perpetrators of that crime, an armed group led by Captain Girolamo Memmo, were still unprosecuted. Bisutti's desecration was in fact a direct attack against the person of Henri IV, the French claimed, and should be treated extremely seriously—not covered up by the complicity of the Venetian legal system, as the murder had been.

Sagredo responded by claiming that he had no jurisdiction over the Gassendi murder case, as the sailors involved were under the command of Memmo, not him. Different accounts of the meaning of Bisutti's vandalism were offered: Bisutti had removed the picture of the French royal family because he thought it was of someone else, or alternatively, that it simply wasn't very good, or that it was poorly printed. He bore no animosity to France or art; in fact, he had a portrait of Henri IV at home, "done by one of the best Parisian painters." Bisutti was the *procuratore* of the church and had removed the picture as being unseemly in a place of worship; the French had not asked his permission to put it up; the priest was in fact in complete agreement with him and had made no statement to the French; and none of the witnesses had been present when he removed the picture, so that all their statements were in fact false.

The murder of Gassendi was also unproven: he died several days after being struck on the head, probably due to medical negligence rather than the blow itself. He had been well known for stealing from foreign merchants, and Bisutti was just one of his many victims. The French themselves knew he was corrupt and were trying to recall him. His death was in fact a favor to them. The whole affair was a manufactured and personal vendetta by the French vice-consul, Louis Savornin.

Deep French animosity in the Levant, claimed the Venetians, had been apparent for years: the French merchants and sailors drastically outnumbered the Venetians in Alexandretta and were trying to make conditions there unbearable, provoking them into minor acts of protest. Now the French were trying to tell them how to run their justice system. The French replied that a character assassination was no justification for a murder, that the Venetians were making Alexandretta unworkable by tacitly condoning criminal activity, that their witness statements were correctly gathered and reliable, and that justice still had to be done.

What is important in this garbled affair is not so much that multiple interpretations are possible and plausible for the historian, but that they are manufactured as such by and for the actors themselves. The central "event," whatever it was, became chameleonically determined by the contexts in which it was situated. The production of rival accounts representing the French and Venetian versions of the event and its relationship to other events was itself part of this process. This is not merely a pragmatic approach of a legal representative attempting to defend a plaintiff, but a deeper ideological commitment to determining intentionality politically. As we have seen in the example of Sagredo's hoax correspondence, such logic could be deployed to produce texts that were simultaneously sincere for one group of readers and satirical for another.

What I would now like to argue is that this logic could also determine the outcome of scientific debates. Soon we will follow Sagredo back to Venice to watch how he brought his political experience of documentary interception and hermeneutic overdetermination to bear on astronomical matters. Such a move was, I think, neither necessary nor unique. The aim of this section is to provide a case study for the further exploration of the transfer of techniques and skills between the realms we now call science and other activities. But first, let us leave Sagredo in Syria and return to Venice with these documentary practices firmly in mind.

Transalpine Messengers

While Sagredo was away trying to manipulate the world from Syria, Galileo finally had his long-awaited break. Compasses, horoscopes, and courtly magnets were set aside— or, rather, reduced into a new re-engineered military instrument, the spyglass. When this failed, Galileo made the instrument astronomical: an Archimedean lever moving the earth closer to the moon. The brilliant little book, barely more than a pamphlet, which narrated its spectacular discoveries ended up being called the *Sidereus nuncius.* Usually read as the herald of modern science, it in fact embodies much of the same subterfuge we witnessed in Sagredo's Palmanova, Venice, and Aleppo. Reapproaching it from this context substantially changes our understanding of the book's publication and initial impact.

The title of the *Sidereus nuncius*—with its ambiguous and even witty play on the two meanings of "nuncius," a message or a messenger—has received much attention.[1] The evidence appealed to in such discussions arises from later attacks on Galileo's intention, and his responses. What has previously gone unnoticed is the extent, lateness, and significance of Galileo's vacillations in choosing his classic title: the book was licensed on 26 February 1610 in Padua as the "Astronomical announcement to astrologers, etc." (*Astronomica denuntiato ad astrologos etc*), the title also recorded in the files sent to the Council of Ten and approved on 1 March. Approval from the coadjutor of the Office against Blasphemy was received only a week later, on 8 March, though no independent documentation survives for this. The book was published less than a week later, and there the printed version of its imprimatur has been changed to "Sydereus nuncius."[2] But there is further evidence for Galileo's indecision regarding his title: Manuscript 70, folio 4 recto of the Galileian Collection in the National Library of Florence contains a note of

the following terms, in Galileo's hand (see plate 4). Each Greek word is followed by its Latin translation and a transliteration: "Κηρυξ, praecox, Ceryx; Μηνυτης, Index, nuntiator, Menytis; Αγγελια, Nuntius, Angelia."[3] These
notes are on the same sheet as some passages from Suetonius and Pliny that
Galileo used in the composition of his dedicatory letter to Cosimo, which
we know occurred very late in the book's composition—probably in early
March, just a week or so before the book was published.

As far as we know, Galileo possessed only one multilingual dictionary
that would help him negotiate between the Latin terms with which he was
familiar and the Greek with which he wanted to perfume his book: Ambrogio Calepino's *Dictionarium*, first published in 1502.[4] It was a standard work,
in fact *the* standard multilingual dictionary for the entire sixteenth century.
There were more than sixty Italian editions of the book in the sixteenth century. In addition, Galileo had a copy of a Greek grammar, Alexander Scot's
Universa grammatica graeca, which included a limited dictionary.[5] It would
seem likely, given Galileo's "little Greek," that he referred to these books
when thinking through the *concetto* of his title.

Not all of Galileo's translations precisely reproduce Calepino and Scot,
and he may have supplemented this resource with others. But the fit is quite
good, and it helps unravel the tangle of meanings around the chosen and rejected terms. Whereas Galileo leads his triads with the Greek terms ("Κηρυξ,"
"Μηνυτης," and "Αγγελια"), following them with their Latin equivalents
("praecox," "Index, nuntiator," and "Nuntius") and transliterations of the
original Greek ("Ceryx," "Menytis," and "Angelia"), we have to work back-
wards from the Latin, presumably as Galileo did.

Calepino's entry for "Praeco" gives translations in Italian ("Trombetti,
banditori"), French ("Crieurs publiques trompettes iurez, qui font les cris
publies") and Spanish ("Pregoneros ò masaieros") and the equivalent Greek
term, "κερικης." The range of meanings shows the legal and oral nature of the
"praeco," semantically very similar to a "herald." Scot's entry for "Κηρυξω"
similarly shows its equivalence to "praeco."[6]

The next entry, "Μηνυτης," is slightly more complicated in that Galileo
offers two Latin terms for one Greek. Scot has no entry for "Μηνυτης"; Calepino has no entry for "Nunciator," only "nunciatio," which he gives as "an
announcement."[7] Calepino's entry for Galileo's first Latin translation of the
term "index," then, shows that Galileo was indeed using this dictionary, as it
refers us precisely to "Μηνυτης." He probably thought of the term "Index,"
then sought its Greek translation. The central use of "Index" for Calepino is
much like the modern English "index"—a guide to finding what one is look-

ing for in a book, or a pointer—but the contemporary vernacular translations are more varied.[8] It is unclear why Galileo would consider "nuntiator" (reporter, announcer) a valid translation of "Μηνυτης," whose semantic range tends more towards the revelation of a secret—or, as a substantive, a legal informer or even a spy. Galileo's choice of "nunciator" may indicate that this was the term whose Greek equivalent he sought. He clearly looked for a term here that conveyed the sense of a messenger, rather than of a message: "nuntiator" rather than "nunciatio." But it seems that he was dissatisfied with such a solution, and continued his search for a more, not less ambiguous term.

Calepino in fact offers the relatively uncommon term "Αγγελια" as a translation of "nunciatio," and this is the last word Galileo explored, which led him to his final choice of "nuncius." For "nuntius" itself, Calepino gives a very precise but nevertheless frustrating entry: "Nuncius, or rather nuntius (Italian: message, messenger, or ambassador; French: messenger, or message; Spanish: messenger or message), αγγελος, αγγελια, the person who announces, and the thing announced."[9] The classical citations that follow are divided to show both the sense of "message" and "messenger." Although "αγγελια" contains a trace of this ambiguity in a way that "αγγελος" does not, its primary sense is clearly "message" rather than "messenger." This may explain why Galileo aborted his search for a Greek noun in his title, and settled for "nuncius" rather than "angelia."

The only reason, then, that the book did not get called something like *Astronomica angelia*, which seems to capture the initial aim of the lexical investigations on the Florentine scrap, is that no Greek term could adequately convey the indeterminacy of "nuncius." The various options sketched out by Galileo are of interest precisely because the terms are not equivalent. They belong to three different semiotic registers: the ceryx or herald is oral/aural, the menytis or informer is gestural, and the angelia or message/messenger is scriptural and visual.

The final choice of "nuncius," which we now know is Galileo's Latinized term for the more theologically loaded Greek "angel," thus signals an active rejection of other modes of conveying information, modes successfully deployed in other astronomical works and therefore not obviously unfit for service. It was perfectly possible for synaesthetizing contemporaries to think of astronomical observation in terms of experiences as varied as conversation, music, painting, or travel. Galileo consciously chose to depict the correct representation of astronomical observation as essentially and only scopic: the *Sidereus nuncius* was to be a visually compelling argument about the potentialities of the visual. The book—small, quickly produced, and relatively cheap—replicated and enacted the instrument it described; it sought to become a paper supplement and simulacrum of the process of telescopic observation.

The *Avviso Astronomico*

Both the genre and the narrative voice through which Galileo chose to convey his arguments were also designated by his choice of title (although the book was already largely written and printed by the time the title was chosen). The "nuncius," which Galileo also translated into Italian as "avviso," was a relatively new genre, especially in print. While variants of the word "nuncius" appeared in books' subtitles with relative frequency in the sixteenth century, they generally functioned to announce the book itself. An "Announcement of health, or Of the incarnation" (*Nuntius salutis, siue De incarnatione*), for example, was printed in Krakow in 1588, but the first systematic use of the word appeared in Germany in 1594, as a subtitle to the biannually spreading European contemporary news collection *Mercurius gallobelgicus*, where it announces political and military information.[10] The popular *Mercurius* was printed, reprinted, pirated, self-pirated, anthologized, translated, and imitated throughout the 1590s and 1600s; one of its spinoffs was published by Latomus, the official printer of the Frankfurt Book Fair catalog, presumably for twice-yearly distribution alongside it. Print publication did not guarantee truth: as John Donne swiftly put it in his early epigram on the new news, *Mercurius Gallo-Belgicus*, "Thy credit lost thy credit."[11] But political news was still heavily commodified in the years preceding the *Sidereus nuncius*; such conditions governed the format and style of the book.

While there is evidence that Galileo toyed with ideas of a divine calling in his role as astronomical explorer, we would do better to consider the *avviso* genre as a narrative tool grasped to transform the snatched glimpses of incoherent data into a seamless, comprehensible prose flow that itself grants identity to the moons of Jupiter. No single observation establishes the existence of the moons: they come into being as scientific objects only as their peculiar, then predictable movements emerge over time. They owe their ontological status to narrative form.

The point of writing an *avviso*, though, is not just that it represents its knowledge as comprehensible, but that it is a politically charged medium. The Venetian case, which tended to retain non-printed *avvisi* longer than other centers of news production, has been well studied by Mario Infelise and Filippo de Vivo. *Avvisi* circulated throughout Europe, containing sensitive political news, rumors, and propaganda. They were collected, recycled, even faked. Roman authorities, in particular, frequently attempted to police the writing of *avvisi*, issuing *bandi* attempting to regulate its writers, who "filled the paper with lies and calumnies."[12]

We can be more specific in providing a thick description of these *avvisi*; during the Interdict, one of the most controversial texts arguing the Venetian case described itself as an *avviso* even though it was nothing like one. Antonio Querini's *Aviso delle ragioni della serenissima Repubblica di Venetia* was published in Venice by Evangelista Deuchino and in Bergamo by Pietro Ventura in Bergamo in 1606. Instead of offering a pseudo-objective description of actual events, its title played on the ambiguities of the word "avviso," which could mean "advice" as well as "view" or "information."[13] De Vivo demonstrates how the printing of Senator Querini's *Aviso* was the result of a change in policy by the Venetian Senate in August 1606. Previously, appeals for its manuscript circulation and printing had been denied, as the leaders favored a policy of refusing to be drawn into a pamphlet war. The publication of Querini's *Aviso* marked an official rejection of this strategy of ostrichism. It was widely perceived as one of the most important expressions of the Venetian position, precisely because its author was a prominent senator. It elicited many responses, and even Bellarmine was drawn into the debate of the *avvisi*, publishing his *Avisi alli sudditi del dominio venetiano* under the pseudonym Matteo Torti in Rome, Bologna, Milan, and Ferrara in 1607.[14] All of these tracts, to be sure, recognized that Querini was not pretending to write an *avviso*, and they, too, similarly offered performative "advice" or "warning" rather than mere description.

Galileo knew Querini well, having discussed natural philosophy with him in the *ridotto Morosini* in the 1590s along with their mutual friend Sarpi. Querini had written to Galileo offering him patronage and protection in 1599, and as one of the rectors of Padua University he signed both the sentence ordering the destruction of Capra's *Usus et fabrica circini* and the imprimatur for Galileo's *Operazioni* in 1606 and *Difesa* in 1607.[15] Querini's *Aviso* and Galileo's *Operazioni* were printed only two months apart. Three years later, when Galileo said that the *Sidereus nuncius* was an *avviso*, it would have been hard for Venetian patricians and the Roman curia not to think of Senator Querini's pamphlet.[16] In a sense, Galileo's *Sidereus nuncius* was an *avviso* precisely in Querini, Possevino, and Bellarmine's sense of the word. It did not merely describe the world; it attempted to change it.

The *Sidereus nuncius* also resembled another popular contemporary genre, the Venetian *relazione*. Even though *avvisi* and *relazioni* were technically separate genres written by and for different audiences, they often circulated together. In its nonofficial use, the term *relazione* might even mean a collection of *avvisi*.[17] Both genres recounted the social, political, cultural, and economic conditions of various peoples around the globe. *Relazioni*, officially secret manuscript versions of diplomats' end-of-mission oral reports, had long been leaked, and had spawned a successful print spinoff, best exemplified by Giovanni Botero's *Relatione universali*.[18]

Relazioni superimposed multiple vectors of differently sourced news so that the discerning reader might spin together a working model of a geopolitical web. As de Vivo has argued, although they are nothing like the objective and exclusively patrician product fetishized by Leopold von Ranke, they offer historians a crucial resource for reconstructing accounts of the nature and flow of information. The Venetian world, more than other comparable information networks, invested in and contributed to the production of *relazioni* and was also increasingly constituted by this media. Diplomats, for example, were given access to archived *relazioni* to prepare themselves for their embassies. Such accounts, at least as much as maps and atlases, structured the world.[19]

Galileo's conceit in calling his book an *avviso* was that it was a cosmic extension of the expanding genre of printed newsletters and *relazioni*. As such, it already made its central argument: that the earth was much like the rest of the cosmos, and not separated from it by metaphysics or physics. Conversely, Jupiter's moons were now part of his information order, brought into existence as scientific objects by his instrument, colonized politically by his language.

Contemporaries' reactions to the *Sidereus nuncius* were split between deep skepticism, enthusiasm, and confusion. These reactions were often based more on how readers felt about Galileo's character than on what he had written or observed. His earlier dispute over the compass with Capra was reignited and fanned from Milan into Germany, creating an anti-Galilean alliance that, already in 1610, extended through some Jesuit networks.[20] One bitter and early reported comment from August 1610 is especially puzzling: In describing the circulation of the sole copy of Martin Horky's anti-Galilean polemic, the *Peregrinatio*, through Germany, Martin Hasdale noted that its first recipient, Marcus Welser, was "completely pro-Spanish." He added: "Do not think that I have spoken out of turn calling him completely Spanish; because the Spanish think, for reason of state, that it is necessary for your book [the *Sidereus nuncius*] to be suppressed as pernicious to religion, with the cloak with which they permit themselves every evil in order to gain total control."[21]

The *Sidereus nuncius* was dangerous to Spain because there they already knew how to read such works: it came from Venice and looked, or at least felt, like an Interdict pamphlet. The Spanish reaction should provoke the historian to ask new questions about this book. What would happen if we were to remove it from its subsequent astronomical history and revisit it in its sites of production? What can it tell us as a material object rather than as a text?[22] Understudied evidence, such as the printer's mark on the title page and incidental woodblock ornaments, will prove crucial in our efforts to understand where the book came from, where it went, and what it might have meant.

Who Printed the *Siderius Nuncius?*

It has universally been assumed that Tommaso Baglioni, whose name appears on the book's title page, was its printer (figure 6.1). In fact Baglioni did not own or run a press until 1615, five years after the appearance of the *Sidereus nuncius.*[23] His appearance on a title page was legally irregular, and it indicates some kind of impropriety.

When we cannot trust the information a book's text supplies, we have to look elsewhere for evidence concerning its production. The first task is to establish the *Sidereus nuncius*'s genealogical relations, as a printed text, to other printed texts, both to uncover the social relations between the texts' producers and to understand the full range of meanings and allusions it bore for contemporary readers. The *Sidereus nuncius* contains four ornamental printing devices, pieces of evidence that may help us understand its production history. The four woodcuts were all recycled from other publications, as was normal. They have, in themselves, no value at all as intended bearers of meaning, and therein lies their historiographical usefulness. As none has adequately had even a bare outline of its history traced, we will take each in turn, and then look at the implications of their combined appearance in the *Sidereus nuncius.*

The most obvious of these woodcuts—and the first to be seen by a reader, though the last to be printed—is the title page's printer's mark (figure 6.1), which I shall call "True Religion" from the phrase inscribed in its border, "HINC RELIGIO VERA" (From Here True Religion).[24] The next woodcut occurs twice, on folios 2 (A2r) and 5 (B1r). The first occurrence, on folio 2 at the head of the dedicatory letter (figure 6.2), was printed last along with the title page, while folio 5 (signature B), the start of the main text, was the first to be printed (figure 6.3). The woodcut is a headpiece with no iconographic significance. We know that the same block was used to print both signatures because there is a chip missing from the lower left corner. Stylistically, the architectural scrolls entwined with garlands are similar to the cartouche surrounding the figure in the printer's mark, but not similar enough to argue

SIDEREVS
NVNCIVS
MAGNA, LONGEQVE ADMIRABILIA

Spectacula pandens, suspiciendaque proponens
vnicuique, præsertim verò

PHILOSOPHIS, atꝗ ASTRONOMIS, quæ à

GALILEO GALILEO
PATRITIO FLORENTINO

Patauini Gymnasij Publico Mathematico

PERSPICILLI

*Nuper à se reperti beneficio sunt obseruata in LVNÆ FACIE, FIXIS IN-
NVMERIS, LACTEO CIRCVLO, STELLIS NEBVLOSIS,
Apprime verò in*

QVATVOR PLANETIS

Circa IOVIS Stellam disparibus interuallis, atque periodis, celeri-
tate mirabili circumuolutis; quos, nemini in hanc vsque
diem cognitos, nouissimè Author depræ-
hendit primus; atque

MEDICEA SIDERA
NVNCVPANDOS DECREVIT.

VENETIIS, Apud Thomam Baglionum. M DC X.

Superiorum Permissu, & Priuilegio.

M VIIII 11. 14.

FIGURE 6.1. The title page of the first edition of Galileo's *Sidereus nuncius* (Venice: 1610), with the "True Religion" printer's mark and Baglioni's name. IC6.G1333.610sa, Houghton Library, Harvard University, title page.

SERENISSIMO
COSMO MEDICES II.
MAGNO HÆTRVRIÆ
DVCI IIII.

*Ræclarum sanè, atque humanitatis
plenum eorum fuit institutum, qui
excellentium virtute virorum res
præclarè gestas ab inuidia tutari,
eorúmque immortalitate digna no-
mina ab obliuione, atque interitu
vindicare conati sunt. Hinc ad memoriam posterita-
tis prodita Imagines, vel marmore insculptæ, vel ex
ære fictæ; hinc positæ Statuæ tam pedestres, quàm
equestres; hinc Columnarum, atque Pyramidum, vt in-
quit ille, sumptus ad Sydera ducti; hinc denique vrbes
ædificatæ, eorúmque insignitæ nominibus, quos grata
posteritas æternitati commendandos existimauit. Eiuf-
modi est enim humanæ mentis conditio, vt nisi assiduis
rerum simulacris in eam extrinsecus irrumpentibus
pulsetur, omnis ex illa recordatio facilè effluat.*

*Verùm alij firmiora, ac diuturniora spectantes, æter-
num summorum virorum præconium non saxis, ac me-*

A 2   *tallis*

FIGURE 6.2. Detail from the *Sidereus nuncius* (Venice: 1610) showing ornamental woodcuts. IC6.
G1333.610sa, Houghton Library, Harvard University, A2r, detail.

ASTRONOMICVS
NVNCIVS

OBSERVATIONES RECENS HABITAS
Noui Perspicilli beneficio in Lunæ facie, Lacteo circulo
Stellisq́ nebulosis, innumeris fixis, necnon in
quatuor Planetis
MEDICEA SYDERA
nuncupatis, nunquam conspectis adhuc cominens,
atque declarans.

AGNA æquidem in hac exiguâ tractatione singulis de Natura speculantibus inspicienda, contemplandaque propono. Magna, inquam, tum ob rei ipsius præstantiam, tum ob inauditam per æuum nouitatem, tum etiam propter Organum, cuius beneficio eadem sensui nostro obuiam sese fecerunt.

Magnum sanè est supra numerosam Inerrantium Stellarum multitudinem, quæ naturali facultate in hunc vsquè diem conspici potuerunt, alias innumeras superaddere, oculisquè palàm exponere, antehac conspectas nunquam, & quæ veteres, ac notas plusquam supra decuplam multiplicitatem superent.

Pulcherrimum, atque visu iocundissimum est, Lunare corpus per sex denas ferè terrestres diametros à nobis remotum, tam ex propinquo intueri, ac si

B   per

FIGURE 6.3. Detail from the *Sidereus nuncius* (Venice: 1610), showing ornamental woodcuts. IC6. G1333.610sa, Houghton Library, Harvard University. B1r, detail.

for a common site of production. Sometimes headpieces incorporated the iconographies of printers' marks, as though to guarantee stylistic continuity; sometimes they were, as in this case, generic pieces, selected according to criteria of size rather than aesthetics, according to the compositor's needs.

Below the headpiece (figure 6.2), at the opening of the dedicatory letter, is an ornamental capital *P*. Elegantly cut, with sharp serifs, it covers a *putto* sitting astride a Renaissance dolphin, seemingly about to spear or whip it with a lily stem. Flowers entwine the couple; the letter has no frame.

On folio 5, at the start of the text proper, another ornamental woodcut capital is placed (figure 6.3). This is from different stock to the putto *P*; it has a double frame and shows or glosses the letter *M*, with a scene of the flaying of Marsyas by Apollo.

The importance of this material lies in the ways it can be used to provide evidence of relationships between the *Sidereus nuncius* and other printed texts. These relationships are at once typographic and social: the reoccurrence of a woodcut can tell us that the same stock, and probably the same press and even perhaps compositor or printer, are at work in different books. Whereas typeface might be used to construct such relationships in early printed books, by the seventeenth century such analysis is all but impossible due to the standardization of fonts. But woodcuts, precisely because they are more prone to physical wear and are unique pieces with a relatively short life span, may be seen as a book's unique distinguishing mark. Whereas other information a book provides about itself—such as its place and date of publication; its publisher's, printer's, or patron's name; and even its title—may all be easily falsified, woodcuts tell us something materially concrete about a book's production.

What, then, can we know about the *Sidereus nuncius* from its woodcuts? The first point, in itself unsurprising but never previously substantiated, is that all four pieces had been used prior to their appearance in the *Sidereus nuncius*. I have not found them all used together elsewhere; only in various combinations. Enough pertinent instances have surfaced, though, to allow a fairly plausible reconstruction of the book's production. This evidence can also be supplemented with other documentary evidence to provide a secure thesis.

P for Putto, Polo

The earliest occurrence of any of the woodcuts' uses is from 1600, when the putto *P* was printed by Niccolò Polo in his Venice edition of Capoleone Ghelfucci's *Il Rosario della Madonna*. The *P* has the same wear we see in its appearance in the *Sidereus nuncius*, making the identification certain and previous use likely.[25]

There are two further pieces of indirect evidence that tie the Polo press to the *Sidereus nuncius*: as Antonio Favaro noted in his report of the discovery of the Padua university rectors' license for the *Sidereus nuncius*, this is appended to another license granted for Giovanni Bellarino's *Doctrina catechismi Romani*.[26] The document is dated 26 February 1610. The request for a license for what would at the last minute change its name to the *Sidereus nuncius* was added, perhaps hastily, to that for Bellarino's book. We know little about the protocol for licensing books at the level of documentary production. But presumably, when more than one book was granted a license within a single document, the request usually came from a single person or representative. Titles mentioned alongside the *Sidereus nuncius* may therefore have something to tell us about its production. The only subsequent edition of the *Doctrina catechismi Romani* to be published in the Veneto was printed in 1610–11 in a joint venture by a Brescian printer Bartolomeo Fontana and Niccolò Polo.[27] It seems likely, then, that Polo submitted the request for Galileo's book alongside that for Bellarino's, as he prepared to complete the work.

This evidence receives further confirmation by the appearance of Niccolo's brother Girolamo Polo's name in the colophon of another Baglioni imprint from 1610: Garzoni's *Piazza universale*.[28] Our initial question concerning the identity of the printer of the *Sidereus nuncius* has been answered: it was printed by Niccolò Polo.

But actually, while we might have identified the owner of the press upon which the book was printed, we may not have yet identified the person responsible for its printing. Baglioni's name still should not be there, even as a publisher. Venetian title pages do not make the same sharp distinctions as, say, English books of the same period between printer, publisher, bookseller, and patron. Any one or combination of these people may appear on a title page, and while one sometimes finds a clear division of roles akin to the English practice of printing *for* and printing *by*, one often has very little idea from a title page about the material production of a Venetian book.[29]

True Religion

Let us return to the woodcuts: before Tommaso Baglioni gained exclusive use of the "True Religion" title page mark in 1608, which he would then go on to use in the 1610 *Sidereus nuncius*, it was used by several other publishers. The earliest use I have found is from 1606, when Giovanni Battista Pulciano uses it for the play *L'Ortigia*, a tragicomedy by Viviano Viviani. This was printed by Roberto Meietti. In other cases, books originally printed by Meietti (or Meietti and Deuchino) are reissued with the "Religio" mark.[30]

The 1607 edition of Confetti's *Privilegiorum sacrorum ordinum* [. . .] *collectio* did not carry the "Religio" title page device, but was published "Sub Signo Italiae" with the press's identity stated in the colophon as "ex typographia Nicolai Poli." Tommaso Baglioni reissued this with a new title page and preliminary matter in 1610. Previously, Meietti had printed it. It seems that both the "Religio" title page device and the "Sign of Italy" came into being to cover Meietti's identity. The "Religio" device was then transferred, at least on paper, to Meietti's employee Baglioni, whose name appeared exclusively alongside it for several years.[31]

We do not know how Galileo went about choosing his printers. His 1607 *Difesa*, though, provides several pieces of evidence as to the true identity both of its own printers and of those of other works: the book has Baglioni's name and Polo's printer's mark on the title page, and Baglioni's name and Meietti's mark as a colophon. In practical terms this probably means that it was printed by Meietti on Polo's press and published by Baglioni. Before addressing the ramifications of this setup, which becomes more curious the harder one looks at it, it is worth looking briefly at evidence pointing both to the 1606 *Operazioni del compasso* and the 1610 *Sidereus nuncius*.

There is a small and previously unnoticed mystery about the *Operazioni*'s printer, Marinelli: he had previously printed thirty-two editions between 1584 and 1591, all featuring his printer's mark of "Abundance." Then there is a gap until 1598, when he apparently published three books in Venice with the mark "Adam and Eve." The *Operazioni*, whose dedicatory letter is dated 10 July 1606, is then the first book after a further eight-year gap. The sporadic new publications, reverting to the old "Abundance" mark and to Padua as place of publication, embrace strange genres: for example, a speech praising a doctoral thesis defense by the author of another Marinelli publication the same year, the Paduan professor Sebastiano Monticello. Then there is nothing again until 1610: Marinelli's final publication, John Wedderburn's defense of Galileo, the *Quator problematum*. We know from Galileo's account books that printing the *Compasso* cost him dearly, and perhaps a real individual called Marinelli reverted to his former profession of printer to capitalize on in-house/vanity publications from the Paduan University environment. Galileo made out payments to a "Mr. Piero, printer," which might well be an alternative spelling or slip for Pietro Marinelli, although there is no printer recorded with the actual name "Piero."Another option is that the press and/or the mark were revived by another printer, and that Marinelli's name provided a convenient cover. As we shall see, one printer who used multiple pseudonyms was Roberto Meietti.

In addition, a jocular sonnet survives, addressed to Galileo by the head printer of, presumably, the *Difesa*, in acknowledgement of an unspecified gift Galileo had given to the printers: "Sir, I've been told / By the guys in the press / To say thank you / For the gift you gave us [. . .] / And if you need to print / Some other birth from your fine wit [. . .] / We offer our humble services / If you find us worthy to command / Don't pass by our sign."[32] It seems that this is precisely what Galileo did when he next "gave birth from his fine wit": he returned to his former printer with his new manuscript. The *Sidereus nuncius* was produced by the same team as the *Difesa*.

Excommunication

The strange thing about finding Roberto Meietti's mark in Galileo's 1607 book is that Meietti had actually been excommunicated the previous year, and that the crime of associating or printing with him was also punishable by excommunication. The *Difesa* made no effort to hide the fact of Meitti's involvement, but actually advertised it alongside that of Polo, who was also known for his Interdict publications. This was the last time Meietti would leave his mark on a book for nearly a decade; already it was a dangerous act. The edict of excommunication, posted in the "usual places"—on the doors of the basilica of St. Peter's in Rome, on the doors of the Holy Office, and in the Campo de' Fiori—and sent out, in theory, to all Inquisitors, publishers, and booksellers, threatened not only Meietti's soul but also his business (figure 6.4):

> Seeing that Roberto Meietti printer of books in Venice has dared, and still dares, to print pernicious books containing heresies, impieties, and various kinds of errors; the Most Illustrious and Most Reverend Lord Cardinals, wishing to make provision against heretical depravity, lest faithful Christians become infected from reading these errors; order each and every person of whatever status, class, condition and preeminence, that they should not dare to obtain any kind of book from Roberto Meietti, whenever printed or to be printed in the future, as they themselves will receive broad sentence of excommunication, from which they may be freed from the articles of death by none but the Holy Apostolic See, and other punishment may be dealt at will by the same Illustrious Lords.[33]

Any publisher or bookseller dealing in Meietti imprints was himself to suffer excommunication, a fine of five hundred ducats, and other punishment.

EDICTVM

Illuſtriſſimorum & Reuerendiſſimorum Dominorum Cardinalium generalium Inquiſitorum.

VM Robertus Meiettus librorum impreſſor Venetijs auſus ſit, & in dies audeat pernicioſos libros imprimere, continentes hæreſes, impietates, ac diuerſi generis errores; Illuſtriſſimi & Reuerendiſſimi Domini Cardinales contra hæreticam prauitatem generales Inquiſitores prouidere volentes, ne Chriſti fidéles ex eorum lectione erroribus inficiantur; præcipiunt omnibus, & ſingulis perſonis cuiuſcumque ſtatus, gradus, conditionis, & præeminentiæ, ne audeant cuiuſcumque generis libros à d. Roberto Meieto quandocumque impreſſos, aut in futurum imprimendos, emere, ſub excommunicationis latæ ſententiæ eo ipſo incurrenda à qua nonniſi à ſancta Sede Apoſtolica præterquam in mortis articulo abſolui poſſint, alijſq. arbitrio eorumdem Illuſtriſſimorum Dominorum infligendis pœnis. Mandantes omnibus, & ſingulis bibliopolis, ac librorum mercatoribus vbicumque exiſtentibus ne prædictos libros emere, aut iam emptos vendere, neque in materia librorum cum eodem Roberto Meietto contrahere, ſeu comercium habere præſumant ſub eiuſdem excommunicationis, ac etiam quingentorum ducatorum, necnon alijs eorum dem Illuſtriſſimorum Dominorum arbitrio infligendis pœnis. Volentes, vt facta huius Edicti publicatione in Vrbe omnes afficiat, ac ſi vnicuique perſonaliter eſſet intimatum. Ingentes nihilominus locorum Ordinarijs, ſeu hæreticæ prauitatis Inquiſitoribus, vt illos in partibus publicari faciant, cuius tranſumptis etiam impreſſis, & ſigillo Sanctæ Romanæ, aut alterius Inquiſitionis, ſeu perſonæ in dignitate Eccleſiaſtica conſtitutæ muniplena fides vbique locorum in iudicio, & extra adhibeatur. Romæ in generali Congregatione ſanctæ Inquiſitionis die xxx. menſis Octobris, anno à Natiuitate Domini noſtri IESV Chriſti M DC VI. Pontificatus Sanctiſſimi in Chriſto patris, & D. N. Domini Pauli diuina prouidentia Papæ Quinti, anno Secundo.

Quintilianus Adrianus Not.

Natiuitate D. N. IESV Chriſti 1606. Indictione 4. die vero 4. menſe Nouembris, Pontificatus Sanctiſſ. in Chriſto Patris, & D. Pauli Diuina prouidentia Papæ Quinti, Anno eius Secundo. Supradictum Edictum, affixum, & publicatum fuit ad Valuas ædis Petri de Vrbe, & necnon ad Valuas palatij ſanctiſſ. Inquiſitionis, ac in acie Campi Floræ, vt moris eſt, per me Io. Baptochium eiuſdem S. D. N. Papæ ac ſanctiſſ. Inquiſitionis Curſorem.

ROMAE
Ex Typographia Vaticana.   M DC VI.

FIGURE 6.4. Edict of excommunication of Roberto Meietti, 1606. Editti e Bandi, Per. Est., 18.4, f.77b, Biblioteca Casanatense, Rome.

All business dealings with Meietti were prohibited. Meietti's books were seized and burned in Rome. No good Catholic was supposed to read any book ever printed by Meietti (this was modified in November 1606 to permit trade in books printed before the Interdict).[34]

From 1606, Roberto Meietti almost exclusively published political tracts defending the Venetian cause in the Interdict crisis. He published all of the most important books and pamphlets arguing Venice's case in its dispute with Rome, on various presses, initially under his own name, then both anonymously and pseudonymously: Sarpi's *Apologia*, *Considerationi*, and *Trattato*, Micanzio's *Confirmatione*, Marsilio's *Difesa*, Manfredi's *Lettere*, and Marsilio's *Seconda parte dell'essame*: these are the most probing and high-risk contributions from Venice and they were all published by Meietti, with his name on the title pages. Many other titles also came out with his name, as well as four under the fictitious imprint "Nicolò Padovano" and at least six anonymously.[35]

The Meietti bookstore, with its sign matching that of the printer's mark, served during the Interdict as an intellectual center for pro-Venetian authors. After the Interdict it also functioned as a meeting place: in 1630, Gian Camillo Glorioso recalled having met Agostino da Mula there in 1610 after Galileo's departure to Florence; da Mula set him a geometrical problem in the bookshop to test him in his candidature as successor to Galileo.[36] Other meetings were less polite: one anonymous denunciation recounts meeting Marsilio there and confronting the author, only to be physically beaten with a copy of Volume Ten of Baronius's *History of the Council of Trent*. Presumably this was the *Annales ecclesiastici* in the 1603 Venetian edition, which at around 775 pages in folio must have hurt. Attempts to bring such authors to the Inquisitors of State resulted in a warning that that the denouncer would not live to see the end of the day.[37]

This was not an isolated incident: as early as 1588, when Meietti first set up shop in Venice, the Venetian Inquisitors questioned him to find out why he was shifting large quantities of books to Padua and selling banned authors such as Rabelais and works by Neoplatonists and Hermeticists. Meietti also used the standard technique of substituting false title pages on prohibited books in order to dupe inspectors. In 1594 he was even found to be distributing banned books in Rome. In 1599 Meietti was threatened with a fine, along with Giovanni Battista Ciotti, Francesco de' Franceschi, and the Sessa company, for the importation of banned books from Germany, including the *Magdeburg Centuries* and David Origanus's 1599 Copernican *Ephemerides*.[38]

Meietti not only imported and exported books across the Alps, but also actively sought to produce a transalpine axis for knowledge transfer. Belluzzi's *Nuova inventione di fabricar fortezze* (1598), for example, was one of the first books on fortification to be printed in Italy. Written between 1545 and 1550, it circulated in various manuscript versions, but was not printed until the end of the century.[39] Baglioni himself wrote the dedicatory letter, addressed to a potential patron, the Calvinist Count of Hanau-Münzenberg, Philip Louis II. In that letter, Baglioni claims to have met his patron at Padua, where he had studied for six months in 1595, possibly under Galileo. Back in Frankfurt, which Baglioni visited at least once between 1595 and 1598, the count had introduced Baglioni to his courtiers, and when Baglioni came across a Belluzzi manuscript, it seemed to offer a fine opportunity to renew the northern connection. This military text seems to have been successful: in 1601 Meietti published a follow-up, a compilation of Italian military writings, including a reprint of Giacomo Lanteri's *Delle offese et diffese delle citta et fortezze*, again with a dedicatory letter from Baglioni, this time to the Austrian protestant noble Heinrich Christoph Thornradl. Both works were immediately advertised in the Frankfurt Book Fair catalogs, demonstrating the existence of a perceived Italophone market for military literature north of the Alps.[40] Editors, printers, publishers, and bookmen did not just manually produce and distribute the words of authors; they acted as brokers, contacts, catalysts, and promoters of knowledge in their own right.

In 1602 Meietti took the extraordinary step of advertising ultramontane books that might be found at his store or warehouse in Venice. The catalog was drawn up by Baglioni.[41] While many of these were standard and entirely acceptable editions, including religious commentaries by Jesuits, several controversial books were included, such as the *Oeuvres* of Rabelais, the *Essais* of Montaigne, and twenty-one titles by Lipsius.[42] Some scientific works were announced, such as a *Teorica planetarum diversorum*, presumably the 1596 edition of Peurbach's *Theoricae novæ planetarum*, published in Basel, and Gilbert's recently published *De magnete*. This was probably the route through which Galileo and Sagredo obtained their copies, which they discussed late in 1602.

Meietti also worked as the official smuggler of banned books for the Republic during the Interdict: in October 1606, a disguised Meietti factor (perhaps Baglioni) managed to slip through the archbishop of Trent's roadblock, but his books were confiscated and taken away for analysis. The factor was attempting to import several thousand Italian and Latin books that Meietti had arranged to have printed over the Alps, as well as a few titles in support of Venice printed in France, Germany, and Switzerland, probably with a view to reprinting them in Venice or passing them on to Sarpi for consideration.[43] Immediately after the seizure of these books on 30 October 1606, Meietti was excommunicated.

Meietti was protected by high-ranking Venetian patronage. In November 1607, with Cardinal Borghese still irate over what had been found in Meietti's barrels of smuggled books, describing him as a man "more worthy of the fire than of the favors of a Prince," the Roman court received weekly intercessions for this "awful man" (*pessimo huomo*) by the Venetian ambassador.[44] We would be wrong to assume that Meietti's publications accurately represent his deepest political convictions. Meietti may have seen in the Interdict an opportunity to escape from the harsh rhythms and low profits of international import and export and unscrupulous reissuing with the potential expansion of swiftly selling pamphlets and government-subsidized tracts.

The little we know of Tommaso Baglioni's career before he became a successful printer and publisher in his own right later in the second decade of the seventeenth century shows that he was always described as Roberto Meietti's "factor"—something between a warehouseman, a salesman, a workman, and an apprentice. He was initially employed by Meietti in 1594, aged eighteen. Occasionally his name would appear on the title page of a book, but this was always in close association with Meietti. Baglioni was dormant as a publisher from 1601 until the outbreak of the Interdict.

Baglioni emerged as a publishing name only as Meietti began to disappear. This is no coincidence. His new publications were political pamphlets. The first, an *Oratione* by Rocco Costantini (dated 25 October 1606), presents the complaints of the inhabitants of Cadore on forest rights, with a Minerva-Lion mark, probably printed by Polo, or at least printed on his presses. The second, the Ludovico Avosto *Oratione* (1607) from "Ambasciator della Città di Bergamo," Baglioni's home town, congratulates Leonardo Donà on his election as doge. This is a genre dominated by Meietti with more than 60 percent of such pamphlets.[45]

In 1611 the rectors of the University of Padua drew up a list of booksellers who were to be notified of the banning of antiroyalist books. Roberto Meietti and Tommaso Baglioni were both included; what is interesting is that still, even after the printing of the *Sidereus nuncius*, Baglioni was referred to as the "agent of Roberto Meietti at the workshop of the two cocks."[46]

In 1613, Baglioni, still without his own press, published Cremonini's *Disputatio*, and we have concrete evidence as to the identity of the printer: in a letter from Lorenzo Pignoria to Galileo a postscript casually reads "A third of Cremonini's book might be missing for it to be finished printing, as Meietti's people tell me."[47] The heterodox book, banned in 1621, describes its publication site only as "apud Thomam Balionum," but was in fact printed by Meietti and published by Baglioni on the Polo presses, using the same ornamental letters as the *Sidereus nuncius*'s putto *P*. Cremonini's next publications, one on philosophy, the *Apologia dictorum Aristotelis*, and one official oration, the *Oratione al sereniss. prencipe Giouanni Bembo*, were both published by Meietti under his own name, after the revocation of his excommunication, in 1616. In 1614, before Meietti was absolved, a warning was issued to "Tommaso Baglioni and another of Roberto Meietti's servants" to desist from selling or printing prohibited books.[48] In 1615, Lorenzo Pignoria commented to Paolo Gualdo on Meietti's strong sales of imported books: "Here in Venice the books from the Frankfurt Fair have arrived, and 24 packs of books have been sold in 8 days by Meietti, for nearly 600 ducats."[49] By 1616, Baglioni had probably set up his own separate business: Pignoria asked Gualdo in January of that year to seek out a Parisian book from "Ciotti, Mr. Tommaso Baglioni, or someone else."[50] Baglioni and Meietti were still on good terms fifteen years later: Baglioni left money to his former boss in his will. Meietti, deprived of political patrons, ended up in financial and philosophical trouble: he was investigated by the Inquisition again in 1621.[51] The *DBI* gives his death as 1634, but in 1646 he was denounced for peddling magical manuscripts copied out for him by monks a decade earlier.[52] From the 1670s to the 1690s, "Roberto Meietti" imprints of Sarpi appeared. It is not clear whether they were printed by his son, as the *DBI* entry suggests, or, more likely, that his name had become a label capable of covering all kinds of illicit printing.[53]

It is impossible that in 1607 Galileo was unaware of Meietti's situation: his information networks extended to Rome, his Paduan colleagues and friends were extremely well informed of the progress of the conflict, and he was fully conversant with both the Venetian and Paduan book trades. Around 1605 he had lent Meietti two copies of Gualterotti's *Poems*, and would certainly have frequented his bookstore in Venice, given Meietti's strength in international scientific titles.[54] Despite his showing extreme reserve in discussing the Interdict in his surviving letters, Galileo's support for the Republic has generally been surmised from his close relationship with Sarpi and other ideologues among the Venetian political elite.[55] In choosing Baglioni, Meietti, and Polo to publish the *Difesa*, Galileo made a conscious show of his loyalty to the Republic. This made good sense: Galileo was defending his reputation against a foreigner, and he used the name of Padua University to do it. Capra's strong Milanese identity probably made him a relatively easy target for a Meietti publication even if the work was not explicitly political but mathematical: Milan had become an important site of production for anti-Venetian tracts. The vituperative rhetoric with which Galileo mocked Capra, the conspicuous

rallying of Venetian and Paduan intellectual and social elites, and even the
title of the work all clearly drew on resources constructed during the barely
concluded Interdict. The very genre Galileo used had been transformed by
the "war of writing": while "defences" had been quite common in literary
criticism before 1606, in that year and the next all four works with that title
published in Italy were pro-Venetian political tracts.

The *Difesa* was born from the Interdict conflict, using the resources of
patronage, rhetoric, and print that were mobilized there. The world of the
Sidereus nuncius might seem far distant from this relatively minor dispute
between a professor and former pupil, but these same resources in fact governed not only its material form, but its choice of genre, language, readership, publication date, and even, to some extent, scientific content.

Strategies

Why, though, would Galileo take such an extraordinary risk? There is nothing
to suggest that he was forced to choose Meietti for the *Difesa* or the *Sidereus
nuncius*: he printed not once but twice, with an excommunicated printer who
was under constant Inquisition surveillance. Given that there were numerous
printers in Padua and Venice who were more than capable of taking on the
modest typographical challenges of the *Sidereus*, why then did Galileo again
choose Meietti? We must remember that the *Sidereus nuncius*'s transformative work on the cosmos and the author, so attentively analyzed by Mario
Biagioli, was not at all inevitable during the book's production. Galileo had
dedicated one book, the *Difesa*, to the Medici and not become their court
philosopher. His Florentine bid in 1610 was clearly stronger than his earlier
efforts with magnets and compasses, but a return to Florence was not and
could not be his only strategy. The choice of Meietti as printer in 1610 must
be seen as a contradictory strategy which Galileo carefully constructed and
then finally rejected: by using a high-risk printer, he made a crucial show of
loyalty to the Venetian elite who continued to profit from Meietti's services
and offer him protection. The traces left by an unnamed Meietti would have
been legible to this reading community.

Furthermore, the choice was not only between a Venetian status quo and
Tuscan transfer: Meietti also opened up other resources and potential career options in a way other printers could not. His privileged access to the
Frankfurt market allowed the *Sidereus nuncius* an immediate and dramatic
entry onto the Northern European stage. The *Sidereus nuncius* was advertised
in the catalogue for the Frankfurt Book Fair less than a fortnight after its

Venetian publication, in an uncharacteristically full announcement Biagioli rightly describes as "bombastic."[56]

The Spring Fair generally ran from the fifth Sunday of Lent to the Tuesday after Easter, which in 1610 would be from 28 March to 13 April.[57] The *Sidereus nuncius* was published on 15 March. We do not know much about the actual use of fair catalogs, but presumably they had to be printed before the fair closed to be of much use to dealers.[58] Sardella calculates the average trade time from Venice to Augsburg in the sixteenth century to be twelve days.[59] Frankfurt would be at least a couple of days more than this. Ottavio Cotogna's *Compendio delle poste* reckons about fifty-five to sixty *poste* from Venice to Frankfurt, depending on one's route. Each *posta* he estimates as between seven and eight miles, so for the journey to take less than a fortnight, the merchandise would have to average well over thirty miles a day, including an Alpine traverse in March. So, for even a handwritten version of the title page of the *Sidereus* to reach Frankfurt in time for its details to be incorporated into the catalog, not a single day must have been lost. Mock-up title pages were sometimes printed and sent to colleagues to advertise a work under press,[60] but this could not be the case with Galileo's *Sidereus*, where the final title was decided upon only during the preparation of material for the dedicatory letter, which was dated 12 March and probably written only a few days previously. Even if Galileo had been able to commit to a definitive title a week earlier and have the book couriered to Frankfurt as a manuscript, he would still have been cutting things very close.

Paul Needham has found in his census of the *Sidereus* that most copies that initially traversed the Alps did not receive the same attention as Italian copies. Northern copies tended to lack the paste-down correction changing "Cosmica" to "Medicea" on folio 5. Of eighty-three copies recorded by Needham, fifty-one lack the cancel.[61] Presumably, this gives us a very rough guide to estimating the proportion of copies immediately sent on the transalpine book route: two in three. Moreover, if we are to believe Galileo's claim to Vinta that less than a week after its release, all 550 copies on the market had already "gone away" (not, it should be noted, "sold out," as is usually claimed), it may be that the Frankfurt fair was the central destination point for the book—indeed, the fundamental justification for its language, style, and form.

The *Sidereus nuncius* left Venice so swiftly that Galileo had no copies to send even valued correspondents, such as Martin Hasdale.[62] This implies a great deal of haste on the part of the exporter in putting together his consignment for the fair. The speed with which the book was written and printed, often commented upon by historians as evidence of Galileo's anxiety over the possibility of losing his precious priority in astronomical discovery, is rather a function of the strong rhythms of the international book trade. These are, in fact, the same issue: when Galileo apologized to Grand Duke Cosimo for the material poverty of the newly published book, he stressed his need to preempt potential priority disputes purely in terms of publication, not printing.[63] It was not enough to write a book, or even to see it into print; what mattered was adequate publication. Even in the case of the *Sidereus nuncius*, where Galileo is generally depicted as shrouding his discoveries in total secrecy, perhaps even withholding crucial sections of text from the printers until the last possible moment lest they leak, oral and scribal publication still traveled faster than the printed book. Kepler, for instance, noted in his *Dissertatio* that he first heard news of the discoveries from his friend Wackher on 15 March in Prague, the very day of the book's publication in Venice. Kepler explicitly commented upon the credibility of the information system that allowed publication to take place without print: "It was men of the highest reputation, raised far above the foolishness of the masses by their knowledge, seriousness, and courage, who reported such things of Galileo and that, indeed, a book was already at the press and would be arriving with the next couriers."[64]

Scientific authors were well aware of the importance of these publishing rhythms, and wrote their works with such timing in mind. Comments on book fair timings show just how extraordinary the appearance of the *Sidereus nuncius* at the Spring Fair was: in 1608 Magini wrote to Aderbale Minerbio, telling him that for books to be included in the Frankfurt fair, he had to have them ready to send from Bologna at the start of February.[65] In April 1614, Cesi urged Galileo to start work on his predictions for the autumnal periods of the Medici stars, as Roman booksellers readied their stock in May for the Autumn Fair, generally held in September.[66] For Galileo to have his *Sidereus nuncius* included in the Spring Fair must have required not only tight organization but also strong business contacts and good credit with the fair, or at least with the printers of the catalog. Meietti was probably the only printer in Venice with strong enough contacts to publish the work at the Frankfurt Spring Fair when Galileo first realized he was writing a book in January.[67] This consideration may well have dictated the completion date of Galileo's manuscript and the tight chronological limits of the book's observations, which were kept to an absolute minimum. The last recorded observation in the book, 2 March, actually postdates its printed imprimatur (1 March). In theory, no new text was supposed to be added to the work after the submission of the manuscript to the rectors of the University. Their approval was dated 26 February, so they must have received what was supposed to be a definitive manuscript some days before this date. Galileo continued to add observations well after this, in an attempt to make his conclusions on the permanence and regular periodicity of the satellites more credible. The tension between the need to provide adequate astronomical data to convince readers of the nonephemerality of the phenomena and the need to have the book reach Frankfurt in time for the fair is visible in these irregular dates.

The Frankfurt Book Fair dominated European international book distribution in the second half of the sixteenth century, and was rivaled only by the book fairs at Lyons and Leipzig.[68] The entire system of transfer and exchange, upon which the more celebrated structure of the Republic of Letters rested, developed out of a standard medieval trade fair. In Henri Estienne's famous paean to the fair, he makes it clear that not only books but maps, instruments, and natural curiosities were sold alongside the standard fare of international European trade. The strength and longevity of the fair makes it easy to dehistoricise, but its fragility as an institution became apparent with the onset of the Thirty Years' War. Both short-range domestic trade and the international market suffered a collapse in the 1620s that had severe effects on scholarly exchange.

The Context of the *Siderius Nuncius*: Anti-Jesuitry and Erotica

One might suppose that the situation in 1610, with the Interdict several years passed, would be radically different from that of 1607. But despite repeat efforts from Meietti and his powerful supporters to have his excommunication withdrawn, this was not to happen for another four years. The polemics of the Interdict, if anything, increased in this period, with British and French cases grafted onto that of Venice. The Gunpowder Plot and the assassination of Henri IV were both swiftly ascribed to Jesuit machinations. Sarpi continued to write, mainly under pseudonyms, directing his mordant satires increasingly against international Jesuitry.[69] Meietti was still extremely active in these debates, producing the most volatile texts he could find.

Several cases in 1610 involving Meietti received the Inquisitor's attention: the first Italian translation of the notorious French *Anticoton*, supposedly printed by Jean Petit in Lyon, was, according to Inquisitorial reports, actually printed by Meietti in Venice.[70] That text supplies, probably intentionally, very little evidence of the place of its production, with a simple triple fleur-de-lys printer's mark and minimal ornamentation. Paper analysis might well prove it to be Italian in origin.[71] The infamous book with its accusations of Jesuitical justifications of tyrannicide against Father Pierre Coton, S.J., came out soon after the assassination of Henri IV in France. It was, according to the exasperated Venetian Inquisitor, sped through the censorship process by Servite readers. Despite the fact that Paolo Sarpi claimed that post-Interdict censorship was at least as bad as before, the case of the *Anti-Cottone* shows that ways did exist for the Republic to keep the Inquisition away from its books.[72]

Sarpi's general anti-Jesuit stance found new focus with the accusations of the *Anti-Cottone*; whether or not he had a hand in suggesting that it be translated into Italian and printed with a false title page in Venice by Meietti, he certainly did his best to encourage its dissemination in the city. Sarpi read the *Anti-Cottone* in late September 1610; the papal nunzio in Venice first noticed the book in early December, "for sale secretly by the bookseller Meietti to his confidants, with the mark of this printer in Lyon, even though I doubt this because of the paper and Meietti's bad character, and that they're printed here, but that's hard to prove."[73] By late December, Sarpi was commenting on the effect the book had had in Venice.[74] Meietti's extraordinary relationship with the Inquisition is made clear by the papal nuncio: "If it were any other bookseller than Meietti, they might stop selling after talking to me and the Father Inquisitor, but with him you have to go straight to the Congregation of the Holy Office and get it banned." The Venetians delayed discussion of the book's banning by repeatedly failing to turn up to meetings with the Inquisitors, rendering their decisions ineffectual. This temporary solution was then replaced with a more permanent one: the appointment of Niccolò Sagredo, Gianfrancesco's father and Galileo's patron in the University of Padua, as assistant to the Holy Office, whose power of veto essentially blocked the Inquisitors' ability to proceed. The case of the *Anti-Cotone* shows us that Meietti was still active as a printer in 1610 and that he took good care to cover his tracks when dealing with dangerous material.

One last case may lead us more directly to those responsible for making the *Sidereus nuncius*. The headpiece of the *Sidereus* also appears in another near contemporary publication, the 1609 *Ducento novelle del Signor Celio Malaspina*. This work, containing titillating gossip and mild erotica, gained notoriety for its banning in 1611. It is important for this story because it is materially linked to the *Sidereus nuncius* in several ways: in addition to reusing the headpiece block, it also contains several letters that appear to come from the same set as the Marsyas *M* (such as an Apollo *A*, Euridice *E*, and Neptune *N*) as well as the "Religio" title page device (figure 6.5). Materially, then, it must have come from Polo's presses. The only information it supplies about its production is on the title page: "In Venetia, MDCIX, Al Segno dell'Italia." This is one of only two uses of the vernacular version of the "Signo Italiae"

DVCENTO
NOVELLE

Del Signor

CELIO MALESPINI,

NELLE QVALI SI RACCONTANO
diuerfi Auuenimenti così lieti, come mefti
& ftrauaganti.

Con tanta copia di fentenze graui, di fcherzi, e motti,

Che non meno fono profitteuoli nella prattica del viuere hu-
mano, che molto grati, e piaceuoli ad vdire.

Con Licenza de' Superiori, & Priuilegio.

IN VENETIA, MDCIX.

Al Segno dell'Italia.

FIGURE 6.5. Title page of Celio Malespina's *Ducento novelle* (Venice: [Meietti], 1609), with the same printer's mark as that which appears on Galileo's *Sidereus nuncius*. Ital 7673.2*, Houghton Library, Harvard University, title page.

that almost certainly masks Meietti. But there is also an independent witness to whose remarks on the book's origin we should listen carefully: the Venetian Inquisitor, whose information in such matters was usually fairly reliable. In a letter dated 3 March 1610 the Inquisitor claimed that "the book of the *Novelle* was printed by Meietti with a false license and name of printer."[75]

The author, Celio Malespina, claimed in a letter to Vinta dated 27 October 1607 that the Inquisitor's imprimatur had already been granted, but it seems likely that this was a lie and that, as the Venetian Inquisitor stated, no such document had ever been issued. The date of the printed book's imprimatur from the heads of the Council of Ten is dated a year later, 17 September 1608, but usually one imprimatur followed the other as a swift formality. The facts of Malespina's biography might help explain the discrepancy. He had spent most of his life as a forger. Those identities and documents that were identified as fake made him, by the late 1570s, persona non grata in Milan, the Holy Roman Empire, Venice, and Florence, with an order for the amputation of his right hand issued in 1579. He returned to Venice in 1579 and tried to turn his crimes into an asset, offering his services as a forger to the doge and the Council of Ten.

Malespina's proposal to the Republic offers a privileged view of the power of documents in Renaissance court culture and the perceived possibilities for that power's subversion. Malespina started his career as a soldier, and he presented forgery as a branch of military strategy no less useful than siege engines and ordnance. Forgery is presented in the petition to the doge as an activity linking machines and machinations: it is a "new invention" that neatly fits into the structures of state building. The potential wonders of forgery are many:

> First, one may cast dissent and discord between Princes, generals, colonels, captains, and other important dignitaries.
>
> In fortresses one may launch many strategies, in times of war and peace, and perhaps easily conquer some.
>
> During a siege one may divert or delay the agreed date of the attack, by putting the general or other officer under suspicion, until relief arrives.
>
> Important prisoners may be freed from prison, victuals liberated, ammunition and artillery lost, and with the artifice of letters make artillery leave the enemy and cut it to ribbons.
>
> In times of financial need, one may obtain a large amount of money in various parts of the world.
>
> One may divert votes towards vacant seats, bring in support from skeptics, and try to pope [i.e., rule the world: *tentare a far il Pontefice*] in one's own way.

One may have anyone anywhere, even far away or under foreign jurisdiction, in one's hands, or at least come to a bad end.

One may disrupt the weddings or marriages of great Princes and other important people, or alternatively profit from them.

In times of famine one may make treaties for grain, wine, oil, and other victuals to provide for one's needs.

One may place infantry and cavalry abroad and pull them out, if need be, from any country.

One may profit from every title of honor or precedence, and just as well do damage.

One may disrupt leagues and peace treaties, and keep messengers everywhere to change letters and ciphers, make permits from any Prince, safe-conducts, letters of credit, passports, and other similar things.

One may at last ruin and discredit all Pashas and other high-ranking officials who serve the Sultan, making them suspected of treason using forged letters, and make them come to a bad end or fall from grace, taking advantage of this opportunity, when they are readying an army to damage Christians divert it (and in this I would very willingly offer myself for the benefit first of God and then of this most happy Dominion), leaving aside for now a thousand other strategies one might respond to, so infinite are the actions of the world, which has always governed itself and continues to do so by means of writing, so that with any of these proposed here, you will always find a way to reap profit or inflict harm, as needs be.[76]

The proposal clearly generated some debate, as the motion upon which the council voted included the clause that Malespina should prove his abilities "both in Latin characters as well as Arabic and code." The motion was defeated twenty-one to two, with two abstentions.

So Malespina went into editing and produced an unauthorized first edition of a purloined manuscript of Tasso's unfinished *Gofreddo* in Venice in 1580. Tasso responded by bringing out his own authorial editions, but Malaspina brilliantly counterattacked with two further editions, each bearing his name in their dedicatory letters, and each actually fuller and better printed than Tasso's own authorized editions.[77]

Malespina then translated from the Spanish Torquemada's *Giardino di fiori curiosi*, which he printed first in 1590 (and reissued in 1591) with Altobello Salicato, who had done his final version of Tasso's text, and then in 1597 (reissued in 1600 and 1604) with Ciotti. It may well have been Ciotti who put Malespina in touch with Baglioni and Meietti for the publication of Malespina's final work, the complex web of translation, imitation, plagiarism, and autobiography printed in 1609 as the *Ducento novelle*. The two hundred stories include tales of forgers, imposters, and deceivers, and although the framing narrative sets their telling in a Boccaccian plague retreat in Treviso in 1572, the stories contain details explicitly postdating this. Almost half the stories are stolen from a well-known French collection first published in 1486, but they are woven into a new fabric with strong autobiographical or pseudoautobiographical elements whose seeming authenticity is impossible to test and hard to resist.

Were the Venetian Inquisitor correct in suspecting that Meietti had both concealed his identity as printer and faked their imprimatur of Malespina's stories, then the web of Malespina's deceptions should be extended to involve other actors. Malespina did not merely try to live as a forger or make forgery the subject of his literature: he also produced forgeries as part of the publication process of his work. His publishers were clearly complicit. Meietti's offenses against the Inquisition went beyond the publication of anti-Roman political propaganda and the camouflaging of his activities, to include the falsification of licensing documentation itself.

The *Ducento novelle* were advertised twice at the Frankfurt Book Fair— first at Easter 1609 as we might expect, just after their publication, but then again at Easter 1610. In May 1609, Alessandro Senesi, the Tuscan agent in Bologna, wrote to Vinta expressing his concern that the book had been published, as it contained scandalous details of *fin-de-siecle* Medici life. Senesi was especially surprised that the book had been "so easily allowed to be printed by the Holy Office."[78]

The *Ducento novelle*, with its "Religio" publisher's mark and false imprint, was published alongside the *Sidereus nuncius*. The books traveled to Frankfurt for international distribution together, alongside other related texts: Baglioni's imprints on sale at the fair included the anti-Spanish Alessandro Campiglia's *La rotonda overo delle perturbationi dell'animo* (1609), dedicated to a doge, and Pedro de Medina's mid-sixteenth-century *Arte del navigare* (1609), dedicated to Galileo's fellow astronomer Antonio Santini.[79] Two other books were by the publisher Giorgio Bizzardo, who came into existence in 1609 and disappeared in 1613, perhaps as another Meietti pseudonym. The other Italian vernacular works supplied by Meietti's import business, the "Venetian Society," were a quixotic comic play (Francesco Andreini's *Le bravure del Capitan Spavento*, published by Somasco), and a prayer book. The starry messenger traveled in strange company.

Masks

One of Sagredo's letters from Aleppo mentions the redeployment of his pseudonymous epistolary hoax against Jesuits in Goa. It arrived grimy from its nine-week Mediterranean cruise, and picked up some interesting comments from Galileo on its back once its seal had been broken. Originally sent on 28 October 1609, it is thought to have arrived in Padua by early January. When edited by Antonio Favaro for his edition of Galileo's correspondence, the presence on its envelope of a sketch of "a grouping of the Medici planets" was noted but not reproduced. Recently, Massimo Bucciantini, Michele Camerota, and Franco Giudice have proposed that the back of the envelope might well contain the first depiction of the satellites of Jupiter.[1] What is important here is less the priority claim for the image on the Aleppan envelope than the relationship between the cumulative traces borne by the material object. The sheet of paper with satellite observations constitutes a graphic assemblage of various forms of early modern inscriptions: instrumentally enhanced light from humanly invisible objects is transformed into inky glyphs jotted onto the back of an envelope in Padua, where they join words written in Aleppo on Venetian paper. Two other kinds of marks are also present: a note in Galileo's hand indicating Sagredo's identity as sender, presumably to help with archiving, and a list of four objects which seems simple, but whose meaning is in fact extremely opaque.

The list reads "Small boxes, cash, thin/fine table, mask" (*Scatolini, soldi, tavoletta sottile, maschera*). These items oscillate between the material and immaterial: while "boxes" are substantial enough, probably indicating purpose-built containers for sending new, delicate objects such as telescopes and lenses to potential patrons, and "cash," though strange on a shopping list, seems easy to explain, whether as a debt, a payment, or a stated ambition, the second half of the list is harder to understand. "Tavoletta" could mean a small table, but the adjective "sottile" makes this unlikely. In *Floating Bodies*, written shortly after the *Sidereus nuncius*, "sottili tavolette di legno" are the wood chips Galileo uses to disprove Aristotelian buoyancy theory. "Tavoletta" could possibly mean a printing plate, such as the copper etching plates for the *Sidereus*'s lunar phases. The 1612 Crusca *Vocabolario* cites one instance where "tavoletta" might mean a small board for painting, but usually it retained the sense of a classical wax writing "tablet"–a surface upon which marks might be incised, such as a board for a woodcut. There are two readings of the phrase that would tie it directly to the production of the *Sidereus nuncius*: one as a technique for printing the Jovian satellite diagrams, the other as a telescopic observation aid. It could be a reference to the pieces of woodblock signaled in Galileo's "Observation Notes" for the printing of the satellite diagrams.[2] This occurs during the observations for 15 January, though, as it is at the top of a sheet, unconnected to the observations that follow, it was probably written at a slightly later date: were this to be the case, the list of items on the Sagredo letter might well be read as a related drafting of the intellectual and material tools necessary for the writing of the *Sidereus nuncius*.

The word "tavola" (table) also carried other, less material senses, as a list or other form of tabulation, and indeed it was by drawing up such tables that the identity and periodicity of the four satellites was established. In this sense, "tavoletta sottile" might mean "small-scale grid," a graph template readied for the plotting of the planets in order to determine their natures. Gingerich and Van Helden have noted the early use of telescopic micrometers for the tracking of the Jovian moons:[3] the "fine table" in Galileo's list could also be such a grid, though the usual term for such a form—at least for the quotidian and ready-made object Galileo might have used, such as the fine wire mold used for paper manufacture—would be "forma," "reticella," or "setaccio." The oscillation between material "tablet" and intellectual "tabulation" does not necessarily require a choice: both might be intended, each enabling the other.

This leaves us with a curious problem when we try to understand what the last item in the list, "Mask" (Maschera), might mean. The *Sidereus nuncius* was written during the run-up to Lent: the book's imprimatur was dated 1 March 1610, which was Shrove Tuesday, the first day of Carnival. It is entirely possible that the "mask" on the list was a reminder for Galileo to buy one for the imminent city carnival celebrations. Alternatively, a "maschera" might possibly refer to some kind of masking device for printing images, or even the cardboard diaphragm Galileo placed near the ocular lens of his telescope to stop down the aperture.[4]

But the word might also be an idea, a strategy for the communication of the discoveries. We generally think of the *Sidereus* as a strikingly "modern" production: the facts speak for themselves; they are self-evident. But when we do this, we anachronistically and teleologically impose categories from modern scientific protocol onto a field where they did not and could not exist. We should take seriously what the book tells us about itself in its title: the facts do not speak for themselves, but are made to speak *as* and *by* something else: a starry messenger. This is the book's mask, the rhetorical persona guaranteeing its fiction of objectivity. The "mask" on the list may be the first inkling of what kind of book these new planets might become.

Negotiating Embassies

"An Ambassador is an honest man sent to lye abroad for the Commonwealth," wrote Henry Wotton in an *Album Amicorum*.[5] The diplomat's social status was high, but his credibility often low. Galileo's "nuncius," eventually chosen from among many other potential masks, was not as stable a category as we might like to think. Some contemporaries were scandalized by Galileo's choice, accusing him of appointing himself as an angelic ambassador from heaven, a heretical new prophet.[6] Others sought to metamorphose the angel into a satirical parody of the author. The author of the first printed attack on the *Sidereus nuncius*, Martin Horky, passed to Kepler these moralizing and libelous notes of Galileo as a fallen angel: "balding; skin covered with syphilis sores; broken skull, crazed brain; optic nerves ruptured through over-curious and obsessive anxiety in observing the minutes and seconds around Jupiter; sight, hearing, taste, and touch gone; gouty hands, due to stealthy theft of philosophical and mathematical cash; heart racing, because he sold everyone a celestial fiction; his guts unnaturally swollen because those in the know are no longer amused by him; feet screaming with gout, as he wanders the four corners of the world. Happy and thrice-blessed the doctor who restores the sick Messenger to his former health."[7]

Gouty and syphilitic, this body had too many organs, ruptured by unreliable instruments and ethics. The human could not become angelic merely by strapping on prosthetic devices. Messengers were, as the case of Sagredo in Syria shows, to be disciplined and questioned by the state along with other subjects. They were not superior beings or remotely controlled probes with perfect intelligence, or even simple representatives of an adequately conveyed policy. As political instruments, they broke down or malfunctioned, failed to fulfill their function in a larger machine, and produced unintended consequences. Worse than that, messengers and diplomats might even be impostors—replicants whose identities could only be known by careful testing.

Within Galileo's ambit in the decade running up to the *Sidereus* there were at least three cases of potentially fake diplomats from far-off courts presenting themselves, whose credibility had to be carefully established by expert analysis of documentation, material evidence, and cross-examination. Early modern states famously paid close attention to inventing and policing intercourt protocols; what is less studied is the system by which the identity and authenticity of an embassy was ascertained. False credentials, forged or intercepted documents, disguises, and even invented languages were all produced as necessary supplements to the norms established by international bureaucracy. Problems over the correct means to establish identity were, of course, central to the functioning of early modern states at many levels.[8]

One particularly well-documented incident occurred in Venice in 1609: a man claiming to be a representative of Shah Abbas of Persia had shown up with a list of impounded items he demanded be restituted to his sovereign. The case of Xwāje Ṣafar is illuminating because it highlights the problems of establishing credibility and credit between systems that are only sporadically in contact.[9] Ṣafar arrived shortly after two rival individuals had also claimed diplomatic status in Rome from Shah Abbas: Ahali Guli Beig and Robert Shirley.[10] The arrival of Ṣafar in Italy coincided exactly, though unrelatedly, with Galileo's telescopic observations and drafting of the *Sidereus nuncius*. As the episode also involved Sagredo, though, we might take it as representative of wider concerns across Mediterranean emporia of adequately representing, translating, and brokering interest over long distances: precisely the concerns Galileo also sought to address, to the dissatisfaction of some readers, in his astronomical tract.

Belisario Vinta, the Tuscan secretary of state and also Galileo's central contact with the Medici court, was the first to deal with the problem of the potentially false diplomats. On 5 January 1610 he wrote to the famous linguist Giovanni Battista Vecchietti, who had recently returned from Persia on a mission to collect Persian biblical manuscripts, asking for his opinion of a troublesome document:

An Armenian has arrived here who says he is Christian and attends Mass, and has brought the attached letter in the attached cloth satchel just as it is, without either the wax seal or the gold cover that the other one brought by

Count Robert Shirley had. He affirms that the said letter is from the King of Persia, and we understand also by other means that in addition to intending to deal with the goods and affairs of the Armenians, he asserts himself to be the Ambassador of the aforementioned King of Persia, and using Interpreters he has already chattered with some people about business in Aleppo and signaled that he has to go to Venice and still has to go where you are [Rome], and as there is no one here who can interpret the letter, Their Highnesses have commanded me to send it to you, as I am now doing, and desire that you interpret it as soon as possible [. . .] . We do not want this place to become a den of Alchemist Ambassadors.[11]

A fortnight later, Vecchietti had translated and authenticated Ṣafar's letter and drafted a response from the Medici to take back to Shah Abbas. Ṣafar would be traveling under a Medici passport as a fully accredited diplomat on his way to Venice. His way was prepared by Vinta with explicit instructions to the Tuscan resident in Venice, Giovanni Bartoli, to advance Medici interests in Persia.[12]

Such Persiaphilia was far from isolated, and not merely based on "wonder" or proto-orientalism. Early seventeenth-century Rome created what might be conceived of as a coherent program of Christian Persianism, constructing churches to the early Persian saints and depicting the embassies of the Shirleys in frescoes in the Salone dei Corazzieri at the Palazzo dei Quirinale.[13] Similar politically conflicting interests were expressed by the Venetians, who produced their own iconography.[14] Interest in Persia was simultaneously antiquarian and contemporary, ranging from humanist plans to improve textual scholarship to the implementation of missionary networks, especially by the Carmelites. Economically, good relations with Persia were seen as crucial to bypass the Ottoman monopoly on imported luxury goods from eastern Asia; militarily, the growing Spanish and Portuguese presence in the Indian Ocean made Persia increasingly important for securing land routes to India; religiously, Christian Armenians, and even Persian Sunni elites, seemed to offer good opportunities for evangelical missionaries.

Venice was far better equipped than Florence or Rome to make sense of Persia, with a permanent position of dragoman, or official translator, whose powers by the early seventeenth century included extensive negotiation. Often an Armenian subject, the dragoman performed textual and oral translations as well as providing reports drawn up after interviews with merchants and diplomats.[15] The Venetian archive contains a small section devoted to Persian documents, often containing both the original and the dragoman's translation.

Solutions to the problems of brokering information and goods, translating between cultures, and sending messages safely across long distances were crucial to the functioning of early modern economic and political systems. The problem of ascertaining and assessing identity was perhaps more visible in dealings with unfamiliar diplomats, but was quite widespread. There were no sure tests for detecting imposters, and each case replied upon the improvisation of notions of adequate context. Sir Henry Wotton provides a nice example relating the arrival at his embassy of Stephen Janiculo, one of several contenders to the throne of Moldavia:

> Here is lately arrived a certain pretendant to the princedom of Moravia [sic], who spendeth his Majesty's name very frankly, as being to recover his right under the protection of the Crown of Great Britainie. [. . .] I protest unto your Lordship, at the first I knew not whether I should take it for a species of frenzy or cozenage. But howsoever, perceiving the matter to require no easy answer, I told him his fashion was fair enough to make me believe much of him, but it was not my fashion to believe men upon so small acquaintance; and therefore, if he could not show me some order from the King my master, I must desire him to provide another host. [. . .] Since, I have understood that he is not altogether a counterfeit, in the generality at least of his pretence to the Princedom, though void of all hope both by the way of Poland and Turkey, who have undertaken the backing of several pretendants.[16]

Part of the problem for Wotton was political: that of knowing whether Janiculo had England's backing. Another part was bureaucratic and philosophical: that of knowing whether the man was who he claimed to be. The two problems were not unrelated. The issue of identity had another dimension which no longer exists: Wotton described his quandary as a "carnival accident," and during this period of the year not only did social hierarchies change, but personal identity became more mutable and malleable. Carnival transformed the mask itself, which was generally moralized as a tricky and deceptive form of imitation, into a tool of experimental self-transformation whose harmful effects were dissipated.

Carnival celebrations were allowed for only strictly determined periods, but the printing of its seasonal plays created a permanent carnivalesque genre. Giulio Cesare Croce, who died in 1609, provides some representatively amusing titles: *The banishment, examination and trial of the fraudulent, insolent and prodigious Carnival, with the confession it made before leaving our country. It's been banished for a year, as its betters deemed fit*; *Tragedy in comedy between the fat and lean mouthfuls on Carnival evening. With the lament of Carnival that Lent comes so soon. And Lent's reply. A gallant caprice*; *The trial or examination of Carnival. In which are understood all the tricks, ruses, caprices, whimsies, messes, intrigues, inventions, news, subtleties, stupidities, idiocies, etc., which it has done this year in our city. With the sentence and judgement formed against it.*[17]

Writing under a mask could mean many different things for a seventeenth-century author. In an undated letter by Angelo Grillo, designed "to provoke [a failed correspondent] sweetly into writing" and printed in Venice in 1608, some contemporary meanings are sketched out:

> With too rigorous a silence you have challenged me to unfriendliness. But I, by nature and profession no friend to duels, refuse the invitation, and wish to show myself to be so peaceful in starting to converse, that you show yourself to be a warrior in starting to fall silent when you should have replied, at least to my words, if not to my show of affection and desire to aid and please you. So I will talk, and tell you that this unexpected change seems strange to me. Unless you were masked for carnival, and in true peace wished to wage a fake war with me. If that's how things are, as I would like to believe, I reply that priests don't wear masks, or enter jousts, and that you should try a different game with me that is more gentle and fitting to my profession, and our friendship; and if you want to fight, get hold of your pen, as I do, challenging you to this battlefield of paper. To arms.[18]

The carnivalesque mask should be distinguished from a wider seventeenth-century discourse on masking and dissimulation, so admirably analyzed by Jean-Pierre Cavaillé and Jon Snyder.[19] While the trope of masked writing does enter both politics and philosophy, most notably in the writings of Sarpi and Descartes, natural philosophy does not seem systematically to adopt a rhetoric or practice of masking to defend itself from perceived threats of censorship.[20] It is not enough here to appeal vaguely to the occult spirit of Rosicrucianism or historical conspiracy theories: what we need to study are the specific models and motives employed by natural philosophers in these disputes.

The disputes over the new star of 1604 had resulted in an adversary referring to the author of a book probably written by Galileo as a "certain cunning mask called Alimberto Mauri."[21] The use of such masks was not a temporary aberration from a fixed model of normal scientific authorship, but rather its general condition. The virtuous, credible, transparent onymous author is rather a special case of personification.

We saw earlier how the publication of Querini's *Aviso* played an important role in initiating the Interdict "war of writing." The genre of *avviso* now seems to oscillate wildly between rhetorics of objectivity and extreme subjectivity: views from nowhere and views from the ground. But these are in fact just a range of authorial conceits to guarantee convincing representations and prescriptions for the world. A good example of the kind of effect that might be produced by such fictions is the pro-Venetian *Most Serene and Merciful Avviso of the Apostolic Church of Rome to His Holiness Paul V,* which claimed to be printed in "The Angelic City at the Sign of Saints Peter and Paul with the privilege of the eternal Saviour."[22] It spoke with what it claimed to be the Church's voice against the military machinations of the pope. Readers were well accustomed to evaluate the sincerity and/or ludicity of such utterances, and authors, too, understood the imaginative opportunities offered in the construction of authorial personae.

The voice of the "nuncius," then, should be understood as being underwritten by a mask or personification, even though the work is not pseudonymous. Far from representing the birth of the modern objective scientific voice, the pamphlet is a carnival piece: shocking in its claims, irreverent in its tone, subversive in its cosmology.

Blotting the Sun

In 1610 and 1611 Galileo Galilei and Christoph Scheiner both observed a new celestial phenomenon. Turning their modified versions of the recently invented astronomical and military instrument, the telescope, towards the sun, they saw that rather than embodying the perfect and unchanging globe described and explained by classical astronomers, the luminous sphere's face was flecked with shifting and moving light and dark spots. In a series of semipublic letters addressed to Marcus Welser, the Augsburg banker, antiquarian, and patron who coordinated the debate, each philosopher offered a different account of the phenomenon, claimed priority in making the first observations, and attacked his adversary's position. Soon after, other astronomers offered their own interpretations, dedicated their works to different patrons, and tried to enter the debate. Twenty years later, the two main protagonists were still at odds over the meaning of what they had seen.[23]

In many respects the sunspot debate, as we now think of it, follows a well-established pattern for early modern natural philosophical literary production. Indeed, it could be seen as the paradigmatic example of courtly debate:[24] A disinterested and wealthy patron invites speculation from two experts over a new discovery; the experts respond in gracious prose, countering each other's rhetorical moves and methodological feints in a disciplined and controlled duel. No mention is made of religious or political difference; the Republic of

Letters celebrates its ideological neutrality and freedom of expression. The reputations of the protagonists and their patron are enhanced; the nature of the phenomenon under scrutiny becomes better understood. This certainly seems to be more or less how the central participants, or at least the team made up of Galileo and his supporters, claimed to understand what they were doing. As Mario Biagioli has argued, the appearance of a dispute was in part created by the decision of the Lincei, Galileo's Roman protectors, to reprint Scheiner's letters as an appendix to Galileo's.[25] But there is something slightly peculiar about the debate, a factor addressed by one of the onlookers that somehow skews the standard account.

In addition to the mediating patron and his two natural philosophers, a fourth figure helped orchestrate the debate. Neither a classic patron nor a participant, Sagredo intervened materially and intellectually in the running of the conversation. Sagredo took it upon himself to make manuscript copies of the letters as they traveled through Venice between the protagonists in Ingolstadt and Florence via Welser in Augsburg. This role was not entirely passive, however. In addition to introducing disruptive delays in the supposedly smooth exchanges between the two protagonists, Sagredo politicized the exchange by drawing attention to the implicit political motives of both Welser and Scheiner in constructing their natural philosophy. While the successful functioning of a debate relied upon the establishment and exchange of credit and credibility between the actors, Sagredo called into question the existence of such credit on two counts. First, he knew from contacts among the intellectual scene of Venice and its province that Welser was a renowned pro-Habsburg (and therefore anti-Venetian) sympathizer and propagandist.[26] His correspondence with leading humanist figures reflected this political stance; Jesuits especially became his intellectual confidants. For Sagredo, this position jeopardized Welser's ability to arbitrate in any kind of debate. Second, Galileo's opponent in the sunspot debate, Christoph Scheiner, chose to publish his letters to Welser on the subject under two classical pseudonyms, "Apelles, behind his picture" and the more defensive "Ulysses under Ajax's shield." Scheiner's real identity remained unrevealed and unknown for almost a year.

Scheiner's use of a pseudonym provoked Sagredo into an outburst that shattered the civility of the debate: "The simpleton Apelles wastes time, paper, and ink in stating the obvious: maybe to show to idiots that he's the defender of truth, he oppresses common speech with his indivisible mathematical points to split hairs with people who talk with real substance, and then disputes things which have been demonstrated, and gets it wrong, so sure of himself because of his unknown name, just like the authors of the *Filotheo* and the *Squitinio*. But it's pointless, because we know all too well who's writing and where his affections and interests lie."[27]

Sagredo linked the falsity of Apelles' arguments to the falsity of his identity. This was a common enough attack against pseudonymous publication in the early modern period. But he went further by honing this general point into a more specific political charge: Filotheo or "Giovanni Filotheo da Asti" was one of the well-known pseudonyms for the Jesuit propagandist Antonio Possevino, who had used the name to defend Pope Paul V's recent anti-Venetian interdict. The other text Sagredo used as an example was even more controversial: the recently published *Squitinio* or *Scrutiny of Venetian liberty*, became an international cause celebre. Using intimate knowledge of Venice's ancient history, it offered a scathing analysis and critique of the myth of Venetian liberty.[28]

The *Squitinio* was anonymous, had a fictitious imprint, and still today has no certain author.[29] There are at least five credible contenders, and each attribution shifts the text's motivation and meaning. Sagredo's example of bibliographical dishonesty as a political (anti-Venetian) act in his letter to Welser was extremely pointed: the first figure to emerge from the Republic of Letter's chatrooms as the probable author of the *Squitinio* was Welser himself (who was elected a member of the Accademia della Crusca for his perfect Italian). The attribution to Welser figured prominently in Pierre Gassendi's 1641 biography of the French *érudit* Nicolas Claude Fabri de Peiresc, whom others suspected of being the author.[30] This attribution had first been made publicly in 1618, in Jean Baptiste Legrain's *Decade*, as an example of books that were flattering to the Habsburgs.[31] By the time of the publication of Vincent Placcius's *Dictionary of Anonyma and Pseudonyma* (1708),[32] and despite the protestations of Welser's 1682 biographer, Welser's authorship had almost become a bibliographical fact.[33] Bayle's *Dictionary* article on Welser finds none of the evidence on the authorship of the *Scrutiny* conclusive, saying: "We have here an example that proves that there are books that make a lot of noise and that we attribute to someone, or someone else, without ever discovering for certain who is the true author."

The charge can be traced right back to 1612, the year the book was published, by following its progress through the labyrinths of the Venetian State Archives. Welser himself sent two copies of the book from Augsburg to his brother in Prague, to be forwarded to the papal nuncio and the Spanish ambassador. The Venetian ambassador to the Imperial Court, Girolamo Soranzo, feigning a lack of interest in the anti-Venetian tract, nevertheless managed to borrow a copy and immediately sent it to the State Inquisitors in Venice. He alerted them that in the book "there are strewn many sharp and poisonous barbs against the Republic."[34] They wrote up a synopsis on the book and sent from Soranzo a semi-ciphered request for further information on its possible author and true place of publication.[35] The book was forwarded to the Council of Ten, who sent their own request for further, more detailed, and certain information and in turn forwarded it to the Savi.[36] The ambassador replied that Welser himself was most probably not only the supplier of the book, but its author:

> This Marcus Welser studied at length in Padua, and possesses detailed knowledge of the Republic. He regards himself especially as an historian, and is considered an extremely erudite person; from this, if it is permissible to form any judgment upon such recondite matters, we may believe that this man may have had a hand in the work, as is also the opinion of others. What troubled me most is seeing that whoever wrote it had easy access to manuscript chronicles, and many ancient and modern writings, from which it seems to me that we may conclude that some Venetian, pulling off a tricky maneuver, has played a major role in it.[37]

At least as important as the identity of the author was that of his source, who seemed to have had unrestricted access to Venice's archives.

The book was returned to the Inquisitors of State, and is now perhaps in the Marciana,[38] with the note "Del Sig.r Marco Velsiero Cittad[in]o d'Augusta" inserted on the title page in an early-seventeenth-century scribal hand and containing heavy annotations, perhaps by readers in the governmental departments through which it passed. Bibliographical issues could also be state issues.

And state issues were personal issues. Sagredo's acerbic comments to Welser on Apelles's hidden identity might contain an implicit attack on Welser too, if the government's opinion that Welser was the real author of the *Squitinio* had been leaked to Sagredo, as seems likely. We might contrast Galileo's relative lack of interest in discovering the true identity of Apelles with Sagredo's probing speculations and evaluations. The sunspot debate was orchestrated by Welser, and he made no offer to reveal the participant's name. With this type of patronage-controlled dialogue, there was no other safe route to the source of the sunspot letters. Galileo's Jesuit contacts in Rome did not seem to know who Apelles was; his patrons in the Accademia dei Lincei had no clue, and other mutual friends of Welser had not been let in on the secret, either. Welser was prepared to reveal only that Apelles was modest (an implicit motive for his use of the pseudonym) and that he would welcome an honest discussion with Galileo.

As Galileo's responses to Apelles took shape, he is increasingly referred to as the "finto" Apelles, meaning feigned, but also fake, as though a truer Apelles would have observed and represented the sun differently.[39] He is then identified as a Jesuit, perhaps as a result of Galileo's reading of the text combined with knowledge of Welser's networks and institutional sympathies and a rough idea of the distribution of good telescopes at the time. This was enough information for dialogue, and for a while his pseudonym itself was turned into a pun, as Apelles became the Jesuit without a name or appellation: "the unApelled jesuit" (*l'inapellato gesuita*). Scheiner did not seem to regard the secret of his identity as particularly important; in January 1613 he revealed himself as Apelles to Giovanni Antonio Magini, professor of astronomy at Bologna, and was visible in his student Johann Locher's pro-Apelles doctoral thesis published in the same year (and probably written by Scheiner).[40] But this information did not become common knowledge. It was not until March of that year that the relieved but bemused Lincei members could inform Galileo that Scheiner's identity had been made public by a fellow Jesuit, D'Aguilon, in his famous *Six Books of Optics*, published in Antwerp.[41]

Sagredo's comment to Welser that he actually knew the identity of Scheiner was itself based on garbled bibliographical networking: Sagredo had correctly identified Apelles as a Jesuit, but he thought that D'Aguilon, who had revealed that Scheiner was Apelles, was in fact himself Apelles. He went on to attempt to lure this doubly-false Apelles into his favorite trap for mathematicians, which involved the political and religious implications of imposing an international date line. The paradox was supposed to undo Jesuit aspirations for a global empire, by exposing the absurdity of two priests in contiguous time zones performing different days' rites at the same moment. Sagredo assumed that his letters were being forwarded to D'Aguilon, whereas in fact they were going to the real fake Apelles, Scheiner, who replied to Sagredo without revealing his identity.[42] The system of secrecy that makes pseudonymic publication work can also discipline attempts to undo it.

The Advancement of Science, Masked

We might be justified in asking to what extent Sagredo's position is representative in natural philosophical debates. Studies on the production of credibility in late seventeenth-century English natural philosophy, for example, stress the importance for authors of guaranteeing the transparency of their identities: Shapin's Robert Boyle was able to make credible claims about the natural world not only because he was a "Christian Virtuoso," but also because he convinced others that this was the case. Shapin's insistence on the importance of ethical rhetoric in the case of some of the central practioners of English science is convincing (even if it does not account for, say, the nature of Boyle's first publication, the anonymous *Invitation to Free Communication*, 1655).[43] But does this analysis provide a generally tenable model for the production of natural philosophical knowledge elsewhere?

In his *Disguised Authors* of 1690, Adrien Baillet, more famous as the biographer of Descartes, posed a rhetorical question: "Who would say that an author who takes the liberty of hiding his face wouldn't have plans to hide something else, too? How would we know that the change in name of one we thought honest wouldn't be a mark or a sign of a change of morals or sentiments in the same person?" Presenting himself as a lawyer acting on behalf of pseudonyms (and a critic exercising extreme good taste and learning), he spent the the rest of the book trying to answer questions of motive in employing pseudonyms. Baillet was quick to point out that we should not impose a single model on pseudonyms. Although some, like Protestants trying to trick Catholics into heretical opinions, were inherently untrustworthy, others, such as Jesuit missionaries in Elizabethan England, were merely trying to avoid unjust prosecution while carrying out the work of God. Modesty, prudence, embarrassment, or just having a strange name were all perfectly good excuses for employing a pseudonym. Vanity, impiety, and cunning may, however, also figure. How was the early modern reader supposed to evaluate such texts?

Sagredo himself, as we have seen, used pseudonyms and elaborate fake identities in his anti-Jesuit epistolary satires of 1608 and after. His charge against "Apelles" of the impropriety of pseudonymity was part of a strategy of writing, rather than any general ethical rule. While legislation did exist to prohibit the use of pseudonyms, De Vivo has shown that during the period of the Venetian Interdict alone, there were more than 130 titles in 250 editions produced in a mere eight months.[44] Of these, about one-quarter were either anonymous or pseudonymous, a Venetian pamphlet being twice as likely as a Roman one to fall into this category. Moreover, early modern authorial practice should be understood as a rich field of positions, rather than a stark choice between anonymity, pseudonymity, or onymity.

Continental models for natural philosophical authorship differed drastically from those advocated in Restoration England. The best example, to illustrate both the variety of positions available to natural philosophical authors and the justifications they deployed, is that of Galileo. The meticulous positivist editorial standards of Antonio Favaro in his *Edizione Nazionale* of the works of Galileo not only guaranteed modern scholarship reliable working texts, but also provided excellent and underused introductory textual notes describing the manuscript genealogy of the works. Galileo's role in the composition of his works varied drastically according to his relations with patrons, coauthors, editors, printers, and the texts or natural phenomena under consideration. These arguments on authorship were, for Favaro, mainly of editorial interest in his quest to establish authoritative texts, but they also furnish a massive amount of evidence for the reconstruction of processes of textual production and consumption, along with vivid access to the professional world of scribes, compositors, printers, booksellers, binders, and dealers involved in making early modern books. An investigation of the relationship between variants in manuscript and printed copies of a work makes possible not only a fixed, definitive text but also a narrative of power and negotiation in textual production. Such narratives do not inhabit an ahistorical and metatextual level, but were a major concern of early modern authors in the texts themselves. Justifications of and attacks on differing notions of authorship make up a surprisingly large part of Galileo's natural philosophical debates. Such preoccupations were not merely aesthetic, but concerned the attempt to define adequate positions to secure the authority for making natural philosophical knowledge.

In his literary showpiece of 1623, *Il Saggiatore*, Galileo offered his own theories of correct authorship. Reconstructing a history of his previous print disputes, he sketched out various positions open to authors. Of particular concern to him were a series of books printed in the previous five years and constituting, in his mind, a discrete debate that *Il Saggiatore* was meant to conclude.[45] In his account of the reception of the *Discorso delle comete*, which was published under the name of Galileo's friend Mario Guiducci but was widely suspected as having been written by Galileo himself, Galileo explains that the book had been wrongly attributed to him by Lotario Sarsi (the anagrammatic pseudonym of the Jesuit Orazio Grassi, against whom *Il Saggiatore* was also responding). There is in fact strong evidence to suggest that Galileo did write most of the text, and a copy of the *Discorso* has recently appeared that contains a dedicatory note to Tommaso Stigliani, the future editor of *Il Saggiatore*, in Galileo's hand, signed "The Author."[46] Having publicly denied authorship of the *Discorso*, Galileo shows how he thinks hidden authors should be treated: "Even if the entire *Discourse on Comets* were to have been the work of my own hand (a thing which will be unthinkable to anyone who knows Mario Guiducci), what would Sarsi's limits be, to uncover my face and unmask me with such ardor while I had made it clear that I wanted to remain unknown?"[47]

This etiquette lesson is then turned on Sarsi, the author of the *Astronomi-*

cal and Philosophical Balance, the book whose truth *Il Saggiatore* claims to test and weigh by reprinting it in its entirety and staging a dialogue with it. Galileo and his fellow Lincean academicians had spent a great deal of effort in establishing the identity of "Lotario Sarsi," but *Il Saggiatore* adopts a novel approach to the pseudonym: instead of unmasking his false adversary, it engages with it as though it were a real person, having revealed it to be a badly made mask:

> I thought that frequently those who went about in masks were either base folk who want to be considered upper-class gentlemen beneath the outfit, and for their own ends take on the value of the honor borne by the nobility; or they are gentlemen who put aside, thus unknown, the just respect due to their rank, and render it permissible, as is usual in many Italian cities, to speak freely with all, carrying on together, absolutely delighted that anyone, whoever he may be, may banter and argue with him without showing respect. I believe that those who cover themselves with the mask of Lotario Sarsi are from this second group (if they'd been from the first, it would have been rather tasteless for them to pretend to be greater than they were). I'm also sure that, unknown as he is, he's decided to say things against me which perhaps barefaced he would not have said, so it should seem too serious that, taking advantage of the privilege granted against masks, I might deal freely with him, nor should he or anyone else weigh every word I might say more freely than he may want.[48]

This strategy in turn forces Sarsi, in his defense, to cite himself, Grassi, as his own teacher. This may seem uncannily postmodern to us, but Jesuit teachers often published under their pupils' names (and occasionally vice versa). What is at stake in the discussion of authorship is not just an opportunity to ridicule a rival, but an attempt to claim and defend ethically and socially acceptable authorial positions from which to make reliable statements about the nature of nature.

The extreme position adopted by Galileo in *Il Saggiatore*—of depicting his own book as a impersonal and precise measuring instrument while blasting his opponent with scathing attacks—is a result of the contortions and convolutions adopted by all actors, real and fictional, over the preceding years.

Sarsi dismisses Guiducci's role as "author" because he claims that Galileo properly fulfils this function, and that he would thus rather speak with the "dictator" than with the "consul."[49] Sarsi plays on Guiducci's status as consul of the Florentine Academy to establish a politicized relationship between Galileo and Guiducci. The teacher-pupil or Socratic-Platonic model is subverted by the ambiguities of "dictator" to fix the spoken word as tyrannical.

For Sarsi, Galilean textual production works by the dissemination of the monologue rather than through the social discourse in which Galileo attempts to represent himself. The Grassi-Sarsi relationship, on the other hand, is the authentic master-pupil transformative process (Sarsi refers to his alter ego Grassi as "My Master"), and this is also situated within the supposedly generous context of lectures: the potential parallels between the Academicians and the Jesuits are effaced in this use of a public/private opposition, merely by introducing the ideal of the pedagogic system in opposition to the closed Academy.[50]

In his *Lettera al M. R. P. Tarquinio Galluzzi della Compagnia di Giesù*, published in 1620, Guiducci responds to Sarsi's attack on the "dictator" Galileo on three fronts. The first is historical, setting the record straight on the nature of the relationship between consuls and dictators in the Roman Republic; the second is by way of counterexample, using Scheiner's pseudonym of Apelles to show how the real "copyists" (*copiatori*) are the Jesuit authors; and the third might be termed grammatological. Guiducci offers a provocative precedent for his own model of authorship in the *Discorso*:

> And if during Plato's lifetime he magnified his gratitude toward his master and honored him in his dialogues by always having him sustain and defend the more reasonable side, why must Sarsi mete out disgrace and blame to me for having studiously sought to emulate the divine skill of that great man? Let him not reply that the nature of dialogues is such that most of the time the persons introduced into them have never so much as dreamed of that which they are fated to say, for Plato himself, in his second letter to Dionysius, expressly declares that he has written nothing of his own, and that nothing is to be found in the work of Plato, but that what he wrote and published came from Socrates his master, who was a very brilliant man and illustrious all his life for his virtue as well as his teachings. Nor would it not be a great impertinence and rashness on the part of any man to call Plato a copyist, and to disdain to quarrel with him but to prefer disputing with Socrates as Dictator? And such is my "ingenuous confession" with regard to having copied that Discourse.[51]

Plato's letter, which Guiducci knowingly refers to but does not quote, offers a beautiful example of the inversions and subterfuges available in the debate over authorship. Plato writes that in order to keep his most difficult ideas pure, they must be withheld from the masses. Therefore his doctrines should not be written, but instead memorized. The next argument is still necessarily surprising: "For this reason I myself have never yet written anything on these subjects, and no treatise by Plato exists or will exist, but those which now bear his name belong to a Socrates become fair and young. Fare thee well, and give me credence; and now, to begin with, read this letter over repeatedly and then burn it up."[52]

Delle comete, Guiducci claims in a gesture of authorial control, is both written by him and not written by him; its "truth" is guaranteed by an "absent," speaking Socratic Galileo, but precisely this position must be protected, safeguarded, by a writing which can only "represent" this speech, written by a writer other than Galileo. Socrates/Galileo can only be written and cannnot write; Plato/Guiducci has not written anything. Guiducci becomes the necessary and nonexistent supplement that guarantees Galileo's absent presence. But at the same time, as Guiducci admits, there is a tension inherent to author functions between, for example, the dialogist and the letter writer, the former writing speech into the mouths of his speakers, the latter claiming that he merely "represents." Moreover, the whole discussion in Plato centers on the issue of textual control: the supposed danger of writing consists partly in its uncontrolled dissemination of "admirable and inspired doctrine" to the "uneducated," to whom they will appear "absurd." But this is precisely the charge reversed by Sarsi in his characterization of Galileo as an oracle, mystifying truth rather than clarifying it.

The *Libra*, Sarsi's reply to this challenge, attempts to restate his autonomy—a tricky move, considering that he is only an anagram. One of the results of Sarsi's insistence that he exists and is his own man is that it prevents Grassi from entering the debate. The pseudonym assassinates the author or, rather, suicides him. Grassi struggles against this process by invoking himself as his own (i.e., Sarsi's) teacher, transforming his pseudonymous author into an amanuensis, into the very process of writing, and nothing more. But the operation of anagrammatization has already positioned him within writing, has already revealed that authorial identity cannot rest on a fictional division between speech and writing. Grassi is speech, Sarsi is writing, and the terms have meaning only relationally. Grassi may attempt to divide the two terms, but he can do this only as Sarsi, and only by writing himself into a text authored by Sarsi.

The cometary debate produced little consensus over comets but a great deal of ill will about authorship. The publication of *Il Saggiatore* is often seen as the Rubicon in Galileo's relationship with the Society of Jesus. It is worth remembering that this was not only a local Roman-Florentine dispute: Sagredo sent Galileo a copy of his 1608 Jesuit epistolary hoax in March 1619, and this may have provided a model for Galileo's subsequent approach to anti-Jesuit writing in the wake of the 1616 condemnation of Copernicanism.[53] The transferal of Interdict polemics to natural philosophy is quite plausible: later in 1619, Galileo sent Sagredo Guiducci's *Discorso*, which Sagredo immediately shared with their mutual friends Sarpi, Micanzio, and Agostino Da Mula; Sagredo in return sent Galileo news of other Jesuit cometary publications.[54]

It seems likely that Galileo's choice of genre for his greatest work, the *Dialogo*, was the direct solution to the problems of authority and authorship raised in the comet debate. There, an even more complex machine of displaced intentionality, modulated responsibility, and surreptitious truth claims was constructed. A figure resembling Galileo himself appears within the text as an offstage authority; his presence is felt, too, in the work's leading marginal glosses and highly subjective index. Presumably, the dialogic form was supposed to insulate Galileo from charges of sincerely held heterodox belief, but the Inquisitor's readers did not approach it that way. Galileo's main defense in his trial was his surprise as a reader that the work's pro-Copernican arguments were so convincing. But here the strategies, acrobatics, and experiments ended: even court philosophers can end up in court.

We should be wary of imposing an anachronistically copyrighted and scientific model on early modern natural philosophical authorship. Despite the fact that the *Sidereus nuncius* is often presented as the example of scientific authorship par excellence, with its dramatic tale of individual invention and discovery, Galileo is closer to the playful *mise-en-âbyme* of Cervantes (whose recent two-part best-seller, translated into Italian in 1622 and 1625, he had read) than to a modern scientific author. The immediacy and transparency of the presentation of astronomical observations in the *Sidereus nuncius* (though, notoriously, not the processes by which they were obtained) seems self-evidently, almost ahistorically, objective. As we have seen, Galileo carefully exploited the generic conventions of the political *avviso* or *nuncius* to secure an unstable authority for his text: the ambiguity of the title—at once a message and a messenger, a document purloined from heaven or an angelic utterance—was intentional.

To give but a few more examples: Galileo's *Istoria e dimostrazioni intorno alle macchie solari e loro accidenti* saw both the author and the patron despair at the compositors' repeated and stubborn transformation of the Tuscan text into Roman; *Il Saggiatore* famously saw Galileo appalled at the insertion into his work of a favorable reference to the poetry of Tomasso Stigliani, who happened incidentally to have been appointed editor. This last case is particularly bizarre in that Stigliani challenged Galileo's list of errata as an affront to his honor, and the text's patron, Cesarini, finally had to intervene and produce a definitive list of 136 typos (more or less halfway between Stigliani's 16 and Galileo's 209). Not only were the texts themselves circulated, edited, and rewritten by interested parties, but Galileo even altered the censor's imprimatur to *Il Saggiatore*, fearing that it was too partisan in his own favor.

It is sometimes not clear within a given text quite who wrote what: the "Letter from the Printer" introducing the *Istoria e dimostrazioni* was actually put together by its two patrons, using the artisanal voice to negate the possibility of dispute. So while printers certainly did not produce the authoritative texts their authors desired, neither did authors. Authorship, however conceived, was always a social activity. The name of the author had a complicated relationship to the identity of any individual. Galileo, as we have seen, also coauthored pseudonymously. The opposition, for Galileo as for the later theorist Baillet, is not between pseudonymous authorship as unreliable and nonpseudonymous authorship as trustworthy, but between honest and dishonest motives in authorship.

Science, Intercepted

Fluid and falsified identities, a porous boundary between manuscript and print cultures, ludic satire and political diatribe: this is not the model of scientific authorship or intellectual history we are used to. Sociologists of science have had to insist on a strong congruence between authorial and natural philosophical identity in order to establish a constructivist epistemology, but the underpinning assumption of social constructivism is that the social is based on an unproblematic definition of a collective of discreet individuals. What is missing from this approach is the possibility of seeing the individual itself as constructed, multiple or virtual, historical. The complex and social process of textual reception makes the author, not the act of writing; "literary technology" has in this sense been curiously invisible to historians of science. The result of this ahistoricism is that social constructivism does not go nearly far enough: the effects of authorial presence and the credibility of documentation are constantly created and negotiated by actors, who are themselves constituted by this very process.

The liberation of history from an ahistorical sociology allows us to test new models of authorship and their relationship to truth claims. These must be constructed from a close analysis of the making of meaning in action, as historical case studies. The very success of the linguistic turn in the history of science, situating rhetorics in historical contexts, has led paradoxically to a circular historiography: we seek out the ethical rhetoric of virtuous practitioners and then claim that this is the nature of scientific practice. I am not denying that such case studies have illuminated their subjects, but they are not, nor should they be, a new master narrative. They need to be complemented by examples that do not conform to such models and which test the limits of their applicability. Sagredo—and, via him, Galileo—offer just such an example; they open the way for us to think productively about scientific actors who never existed corporally, about the relationships between authors and other actors involved in the making of scientific texts, and about the reception of scientific texts as part of scientific practice. We may well find that the model we currently employ, with its neat pairings of credit and credibility, virtue and virtuosity, reason and reasonableness, is the anomaly rather than the norm, or even that scientific practice was so localized and nontransferrable that the idea of a norm belongs to an Enlightenment, rather than early modern, epistemological landscape.

This study is intended more as an implicit critique than as a new model, but I hope it might suggest some avenues for future research. The turn towards material history, for example, has often been presented as a move away from the text, but this should not be the case. Close readings must be extended to include the materiality of texts themselves. While this has become normal practice in other fields, the welcome rush by historians of science to grasp at "things" has peculiarly precluded the potentialities of studying the most heavily produced scientific object of the early modern period: the book. Once we recognize that textual relations are social relations, we might be able to see the politics of information orders as actors themselves saw them.

The cult of gentility, first propagated by Republicans of Letters themselves, might also be displaced by studies of debates in action. Institutions have ways of disciplining their members into polite compliance, though we should not overestimate this power. Rather than accepting codes of civility as the accepted rules of the game, we would do well to see how knowledge fares when produced by rude, boisterous, scathing, or curmudgeonly practitioners. We might find that the choice is not starkly set between order and anarchy, but across a spectrum of practices, including satirical sideswipes and grotesque guffaws. We have generally seen the Hookes, Harrisons, and Hobbeses as outsiders and outliers, but this is to adopt an asymmetrical position, tacitly assuming that institutional science was normal science. The closer one looks at even the main exemplars of civil science, the less genteel they seem. Civility was insisted upon so much because it was so lacking, not because it was universally accepted. We like to study scientific disputes because they make epistemological fault lines evident, but we should not therefore assume that there was no dispute about the nature of disputation itself.

There is another sense in which such an approach might bear fruit. Readers' techniques of unmasking, revealing, or attributing authorship to anonyma and pseudonyma were applied as an epistemological model in depicting and interpreting the work of natural philosophy itself: Nature had to be forced to "reveal her secrets" by the sharp-eyed investigator, just as the discerning reader penetrated the occult mysteries of the Republic of Letters. Francis Bacon's Merchants of Light were, after all, spies. A great deal of energy was expended shoring up the protocols that protected scientific data and inventions from prying eyes, but this is less the victory of an "official" epistemology over a parasitic threat than a declaration that some spies were better than others. This aspect of the interpenetration of the history of the book and the history of science remains relatively unexplored.

The fading traditional skills of critical bibliography and archival research have to be applied to newer questions of historical epistemology in order to move both forward. Only by unpacking and historicizing the practices and techniques of the craft of authorship and readership can we begin to understand the workings of their textual products, and begin to understand the production of natural philosophy.

PLATE 1. Sagredo as a young man. *Gianfrancesco Sagredo* by Gerolamo Bassano (before 1612). Zhytomyr Regional Museum, Ukraine.

PLATE 2. Sagredo in diplomatic robes, with a view of ancient Alexandria. *Gianfrancesco Sagredo* by Gerolamo and Leandro Bassano (1619). Ashmolean Museum, University of Oxford.

PLATE 3. Sagredo's Persian kilim and Syrian *commissione dogale. Gianfrancesco Sagredo by* Gerolamo and Leandro Bassano (1619), detail. Ashmolean Museum, University of Oxford.

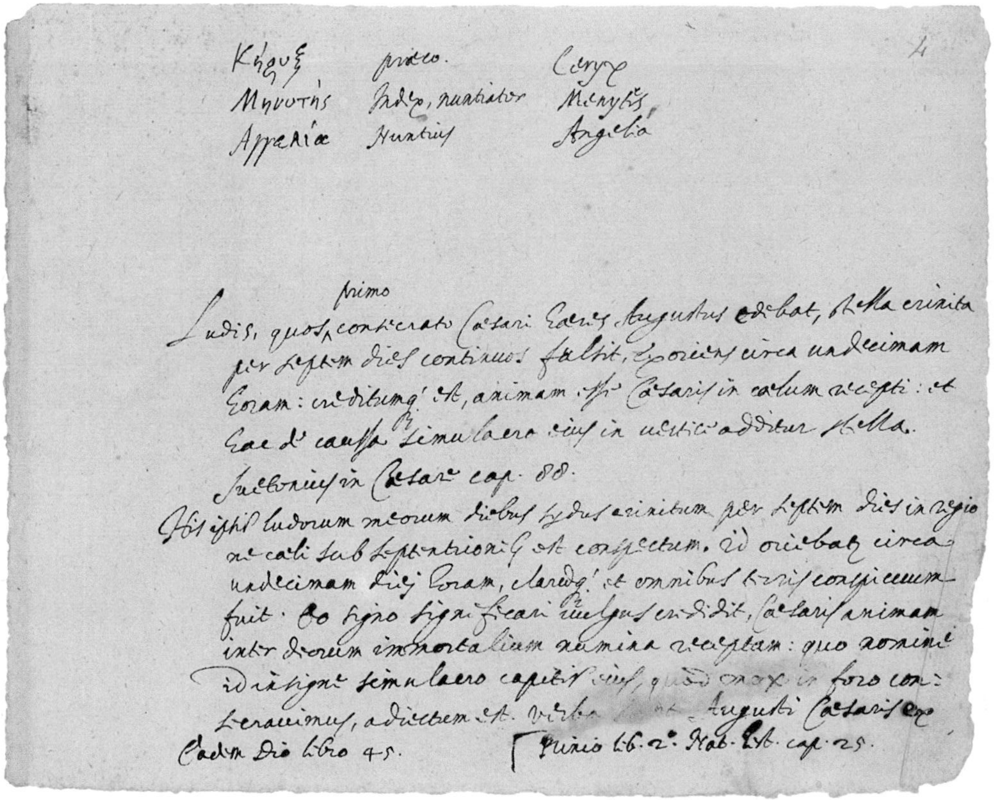

PLATE 4. Galileo's notes documenting alternative titles to the *Sidereus nuncius*. BNCF, Mss. Gal. 70, f.4r.

Acknowledgments

Unlike most first academic books, this is not a revised doctoral thesis. Nevertheless, my first thanks go to the members of my PhD committee at the European University Institute, Florence. John Brewer was an excellent noninterventionist supervisor, always there when needed, never prying. He taught me that history could be interesting, engaged, rigorous, and above all pleasurable. Mario Biagioli first suggested, in a comment on a footnote in the thesis, that the idea of writing a book around Sagredo might be viable. He later confessed that he himself had wanted to do this, but then, dissuaded by Gaetano Cozzi, he moved on to better things. I am sure this is not the book he would have written, but hope it does not disappoint. Paula Findlen provided encouragement, humor, ideas, and my first entry to American academia, bringing with it not only the great gift of a class of students, but also the glimmer of a possibility that I might be able to do this sort of thing professionally one day. Peter Becker was the most humane historian of bureaucracy I can imagine, a model of integrity and high professional standards. Simon Schaffer, who was also my British Academy Post Doctoral Mentor, has been a source of unlimited wisdom, knowledge, and wit to an entire generation of students. Every conversation is a revelation, every pint a pleasure.

The book evolved through what seems in retrospect to be a vast number of institutions, and alongside many other projects. At Stanford and Cambridge I was fortunate to find students and colleagues who introduced me to the history of the book as an integral part of the history of science. The Italian Academy for Advanced Studies in America at Columbia University, under the expert guidance of David Freedberg, provided an exciting and invigorating home. The Department of History at the University of Miami welcomed me warmly, taught me how to teach, and prepared me for a real job. My colleagues in the Department of History at Georgia State University have been model citizens in difficult times, and I deeply appreciate their commitment and engagement. I am extremely grateful, too, for their help in covering my absences during early parenthood. Thanks also go to my students, both undergraduate and graduate, for all those questions, answers, and jokes.

Georgia State University supported this project with a Research Initiation Grant in 2010, and the Department of History also awarded me both an Ellen L. Evans Grant and a Copen Grant. Research was also funded by a Grant for Independent Research on Venetian History and Culture by the Gladys Krieble Delmas Foundation in 2007 and an NEH Post-Doctoral Rome Prize in Renaissance and Early Modern Studies at the American Academy in Rome in 2009–10. The Academy offered a perfect environment for interdisciplinary scholarship: friendship, fine food, and football (i.e., *calcio*). Any views, findings, conclusions, or recommendations expressed in this book do not necessarily represent those of the National Endowment for the Humanities.

I owe a colossal debt to the many archivists and librarians who helped me over the years in Venice, Rome, Florence, Paris, London, Oxford, Cambridge, New York, San Francisco, Berkeley, Atlanta, and elsewhere. Without them, there would be no such thing as historical research or history. I would also like to thank the many groups who hosted various versions of this material: the Museum of the History of Science in Oxford; the Department of History and Philosophy of Science at Cambridge; the University of Warwick in Venice; the British Academy Post Doctoral Fellow Symposium; EMPHASIS (Early Modern Philosophy and the Scientific Imagination) at the School of Advanced Study, University of London; the Italian Academy for Advanced Studies at Columbia University; the History and Philosophy of Science Seminar Series at McGill University; Rider University; the University of Cagliari; the History of Science Colloquium at the Max-Planck-Institute for the History of Science, Berlin; the Department of Italian Studies at the University of California, Berkeley; the Mellon Science and Print Culture Workshop; the University of Wisconsin–Madison; the History of Science Society; and the Renaissance Society of America.

Ideas, information, and aid were generously offered by, among others, Albert Ascoli, Leonard Barkan, Robert Batchelor, Jim Bennett, Francesco Borghesi, Tony Bracewell, Horst Bredekamp, Corey Brennan, Marty Brody, Jon Calame, Michele Camerota, Kathleen Coleman, Jonathan Conant, Lidia Dachnenko, Tom de la Mare, Nick Dew, Brendan Dooley, Caterina Enni, Seth Fagen, Carmella Vircillo Franklin, Paolo Galluzzi, Oscar Ganzina, Roger Gaskell, Owen Gingerich, Jeffrey Glover, Edward Goldberg, Michael John Gorman, Stephen Greenblatt, John Heilbron, Matthew Humberstone, Mario Infelise, Adrian Johns, Henry Johnson, Matthew Jones, Edward Kasinec, Lauren Kassell, Richard Keatley, Hannah Kendall, John Krige, Richard Lan, Mary Lindemann, Henry Lowood, Noel Malcolm, Russell Maret, Massimo Mazzotti, Charlotte Miller, Kiel Moe, Jason Moralee, Asha Nadkarni, Paul Needham, Paul Nelles, Will Noel, Irina Oryshkevich, Katherine Park, Larissa Petrivna, Luca Polese, Joad Raymond, Ingrid Rowland, Patrizia Ruffo, Guido Ruggiero, Neil Safier, Jake Selwood, Pamela Smith, Barbara Spackman, Cristina Stango, Randy Starn, Will Stenhouse, Daniel Stolzenberg, Bill Stoneman, Helen Szépe, Ramie Targoff, Jon Thompson, Lela Urquhart, Matteo Valleriani, Ann Vasaly, Rick Watson, Gillian Weiss, Catherine Whistler, Richard Wittman, and Roberto Zago. My thanks.

Filippo de Vivo, Mario Biagioli, and Eileen Reeves read and commented on several drafts, working quiet miracles, feeding me documents, guiding me towards the light, and saving me from some embarrassment. It goes without saying that all remaining mistakes, misinterpretations, and blunders are solely mine. My editor at the University of Chicago Press, Karen Merikangas Darling, has calmly initiated me into becoming an author, and has trusted in this book for many years. Renaldo Migaldi has been an exemplary textual editor and perceptive critic.

This book is dedicated to Rena Diamond, who has put up with Sagredo for too long. I am grateful beyond words. Without her amazing patience, deep trust, brilliance, and love, this book would never have been finished. Now it is, and it is hers.

Abbreviations

ACDF: Archivio Congregatio pro Doctrina Fidei, Vatican City

ARSI: Archivum Romanum Societatis Iesu, Rome

ASF: Archivio di Stato di Firenze

ASV: Archivio di Stato di Venezia

BNCF: Biblioteca Nazionale Centrale di Firenze

CASANATENSE: Biblioteca Casanatense, Rome

CORRER: Biblioteca del Museo Correr, Venice

CSP, Ven.: *Calendar of State Papers Relating to English Affairs in the Archives of Venice*, edited by Horatio F. Brown. Vol. 9, *1592–1603*. London: Eyre and Spottiswoode, 1897. Vol. 11, *1607–1610*. London: Mackie and Co., 1904. Vol. 12, *1610–1613*. London: Mackie and Co., 1905.

DBI: *Dizionario Biografico degli Italiani* (http://www.treccani.it/biografie/)

LINCEI: Biblioteca dell'Accademia Nazionale dei Lincei e Corsiniana, Rome

MARCIANA: Biblioteca Nazionale Marciana, Venice

ODNB: *Oxford Dictionary of National Biography* (http://www.oxforddnb.com)

OG: Galileo Galilei, *Le Opere di Galileo Galilei*, edited by Antonio Favaro. 20 vols. Florence: Barbera, 1890–1909.

Notes

Introduction

1. On Sagredo, see Antonio Favaro's essential articles: "Giovanfrancesco Sagredo," in *Amici e corrispondenti di Galileo*, ed. Paolo Galluzzi (Florence: Libreria editrice Salimbeni, 1983); "Ancora a proposito di Giovanfrancesco Sagredo" in *Scampoli galileiani*, ed. Lucia Rossetti and Maria Laura Soppelsa, vol. 2 (Trieste: Lint, 1992); "Giovanfrancesco Sagredo e Guglielmo Gilbert" in *Adversaria galilaeiana: Serie I–VII*, ed. Lucia Rossetti and Maria Laura Soppelsa, (Trieste: Lint, 1992) and also chapter XX of Favaro's *Galileo Galilei e lo studio di Padova* (Padua: Antenore, 1966). Galileo's letters to Sagredo are lost; there are 102 from Sagredo to Galileo printed in Galileo Galilei, *Le opere di Galileo Galilei* (henceforth *OG*). See also Matteo Valleriani, *Galileo Engineer* (Dordrecht: Springer, 2010), which translates many of Sagredo's letters to Galileo concerning mechanics and artisinal epistemology, and my "Galileo's Idol: Gianfrancesco Sagredo Unveiled," *Galilaeana: Journal of Galilean Studies* (2006): 3.

2. Valleriani, *Galileo Engineer*, provides a good introduction to many aspects of Sagredo's intellectual biography.

3. See, for example, Owen Hannaway, "Laboratory Design and the Aim of Science: Andreas Libavius versus Tycho Brahe," *Isis* 77 (1986); and Steven Shapin, "'The Mind Is Its Own Place': Science and Solitude in Seventeenth-Century England," *Science in Context* 4 (1991).

4. See, most informatively, the new introduction to the reprint of Simon Schaffer and Steven Shapin's *Leviathan and the Air-Pump: Hobbes, Boyle, and the Experimental Life* (Princeton, NJ: Princeton University Press, 2011).

5. The impetus for this was another strong case study of Restoration London: Adrian Johns, *The Nature of the Book: Print and Knowledge in the Making* (Chicago: University of Chicago Press, 1998).

6. Elizabeth Yale's "Marginalia, Commonplaces, and Correspondence: Scribal Exchange in Early Modern Science," *Studies in History and Philosophy of Biological and Biomedical Sciences* 42 (2011), offers a good point of entry for the English scientific manuscript scene; Brian Richardson's *Manuscript Culture in Renaissance Italy* (Cambridge: Cambridge University Press, 2009) is the best general survey of scribal publication.

7. The classic study is Mario Biagioli's *Galileo Courtier: The Practice of Science in the Culture of Absolutism* (Chicago: University of Chicago Press, 1993).

8. Robert Batchelor's London: *The Selden Map and the Making of a Global City* (Chicago: University of Chicago Press, 2013) liberates London from its Atlanticist confines.

9. See Avner Ben-Zaken's *Cross-Cultural Scientific Exchanges in the Eastern Mediterranean, 1560–1660* (Baltimore: Johns Hopkins University Press, 2010) for a welcome attempt to reconstruct Mediterranean science.

Chapter One

1. *OG* 14:350, 355; 15:31.

2. *OG* 7:31 (translation mine). The best critical edition is Galileo Galilei, *Dialogo sopra i due massimi sistemi del mondo tolomaico e copernicano*, ed. Ottavio Besomi and Mario Helbing. 2 vols. (Padua: Antenore, 1998).

3. *OG* 7:197, 466.

4. On this episode see Gaetano Cozzi, *Paolo Sarpi tra Venezia e l'Europa* (Turin: Einaudi, 1978), 175, and Sabina Pavone, *Le Astuzie dei Gesuiti: Le false Istruzioni Segrete della Compagnia di Gesù e la Polemica Antigesuita nei secoli XVII e XVIII* (Rome, Salerno, 2000), 161–67, as well as chapter 6 in this volume.

5. *OG* 19:639.

6. Marco Foscarini, *Della letteratura veneziana ed altri scritti intorno ad essa*, (Bologna: Forni, 1976), reprint of 1854 edn. (1st edn. 1752), 337; quoted in Favaro, "Giovanfrancesco Sagredo," 201.

7. *OG* 12:45–46. The rest of the letter is less noble in sentiment, accusing his Jesuit opponent "Apelles" of mathematical incompetence, scheming, and dogmatism.

8. Biagioli, *Galileo, Courtier*, 251–52.

9. On the burgeoning history of friendship, see, most importantly, Alan Bray, *The Friend* (Chicago: University of Chicago Press, 2003). More specifically, for case studies of the politics of ideal friendship in Venice, see Gaetano Cozzi, "Una vicenda della Venezia barocca: Marco Trevisan e la sua eroica amicizia," in *Venezia barocca: Conflitti di uomini e idee nella crisi del Seicento veneziano* (Venice: Il Cardo, 1995), originally published in *Bollettino dell'Istituto di Storia della Società e dello Stato, Fondazione Giorgio Cini* 2 (1960). See also Peter Miller, "Friendship and Conversation in Seventeenth-Century Venice," *Journal of Modern History* 73 (2001): 1–31. Both essays mention Sarpi's reading of Montaigne's essay on friendship.

10. Guido Casoni, *Ode dell'illust. et eccell. signore Guido Casoni dedicate all'illustriss. & reuerendiss. sig. cardinale Cinthio Aldobrandini* (Venice: Gio. Battista Ciotti, 1601) (2nd edn.), 111. Casoni was one of Sagredo's few traceable clients: the *Ode* includes poems to Gianfrancesco's father, Nicolò (p. 64, described as a great admirer of Tasso), and brother Zaccaria (lover of Petrarch). Its social world includes Carlo Belengo's informal literary academy, the art collector Carlo Ruzini, and a Tintoretto painting. In 1598 Casoni produced the family's single effort at religious propaganda, an account of their martyred ancestor St. Gerardo Sagredo: *Vita del glorioso santo Gerardo Sagredo nobile venetiano, monaco dell'ordine di san Benedetto* (Venice: Domenico Nicolini, 1598).

11. For excellent discussions of early modern skepticism concerning vision, see Stuart Clark's *Vanities of the Eye* (Oxford: Oxford University Press, 2007) and Joanna Picciotto's *Labors of Innocence in Early Modern England* (Cambridge, MA: Harvard University Press, 2010). It is worth remembering that Monteverdi's *Orfeo*, with its climactic punishment of prohibited vision, "O dolcissimi lumi, io pur vi veggio / io pur . . . ma qual eclissi, ohimè, v'oscura?" was first performed in Mantua in 1607.

12. *OG* 16:414. Despite the 1612 Crusca dictionary's definition of "idolo" purely in terms of idolatry, writers such as Dante, Petrarch, and Ariosto, all of whom Galileo was familiar with, used the term to mean either a mental image or a beloved object. Galileo uses the word only once in Italian, and once uses "idola" in Latin, where it corresponds to "imago": *OG* 8:627 (an undated fragment, presumably from the 1610s). For a good introduction to early modern idols and idolatry, see the volume of *Journal of the History of Ideas* devoted to the subject, especially Jonathan Sheehan, "Introduction: Thinking about Idols in Early Modern Europe," *Journal of the History of Ideas* 67 (2006): 561–70.

13. On Galileo's portrait, which was sent to Sagredo in June 1619, see Favaro, "Studi e ricerche per una iconografia galileiana," *Atti del Reale Istituto Veneto di Scienze, Lettere ed Arti*, 72 (1913): 1003. Galileo's portrait was not, however, listed amongst Sagredo's possessions at his death. See note 60. Sagredo refers to the exchange of portraits as part of a long series of demonstrations of their "antica, sincera, reciproca et incorporabile amicitia" *OG* 12:464.

14. Favaro, "Inventario della eredità di Galileo," in *Scampoli Galileiani* 1:64–69.

15. Foscarini, *Della letteratura veneziana*, 316–17 (cited in Favaro, "*Giovanfrancesco Sagredo*" 265n1). A letter from Foscarini to Cocchi dated 16 January 1744/45, listed in *Le Carte di Antonio Cocchi: Inventario*, edited by Anna Maria Megale Valenti (Florence: Giunta Regionale Toscana, 1990), 20n444, and conserved in the Archivio Baldasseroni in Florence, thanks Cocchi for the portrait.

16. "Nota de' ritratti di uomini illustri che sono presso il Senatore Conte Capponi," BNCF, Ms. II.-.184, 29ff187–89 (Getty # I-3024).

17. On Gerolamo (1566–1621) and Leandro (1557–1622) Bassano, see (under *Dal Ponte*) *DBI*; Giuseppe Gerola, *Bassano* (Bergamo: Istituto italiano d'arti grafiche, 1910), 114–21, and Edoardo Arslan, *I Bassano* (Milan: Ceschina, 1960), 2 vols., 1:257–92.

18. Larissa Salmina-Haskell, review of *Kartini Italianskih Masterov XIV–XVIII Vekov is Museev SSSR* (*Картены Итальянских мастеров XIV–XVIII веков из Музеев СССР* [Moscow: Советский художник, 1986]) by Viktoria Markova, *Burlington Magazine* 133 (1991): 630.

19. L. N. Sak and O. K. Shkolyarenko, eds., *Західноєвропейський живопис* (West-European Painting, 14th–18th Centuries) (Kiev: *Мистецтвоб*, 1981), plate 20.

20. For a history of the Sagredo collection that briefly discusses Gianfrancesco but fails to notice his correspondence with Galileo on the Bassani, see Cristiana Mazza, *I Sagredo: Committenti e collezionisti d'arte nella Venezia del Sei e Settecento* (Venice: Istituto veneto di scienze, lettere ed arti, 2004). Other studies on the later Sagredo collection that similarly ignore Gianfrancesco's contribution include Francis Haskell, *Patrons and Painters: A Study in the Relations between Italian Art and Society in the Age of the Baroque*, 2nd ed. (New Haven: Yale University Press, 1980), 263–67; Alice Binion, "Algarotti's Sagredo Inventory," *Master Drawings* 21 (1983); Cristiana Mazza, "Frammenti inediti della scomparsa pinacoteca Sagredo," *Arte in Friuli. Arte a Trieste* 15 (1995); Cristiana Mazza, "La committenza artistica del futuro doge Nicolò Sagredo e l'inventario di Agostino Lama," *Arte veneta* 51 (1997); Linda Borean, "'In camera dove dormo': Su alcuni quadri di Nicolò Sagredo," *Arte veneta* 50 (1997), and William R. Rearick, "More Veronese Drawings from the Sagredo Collection," *Master Drawings* 33, (1995): 132–43. The Sagredo collection of Nicolò, Gianfrancesco's nephew, was described by Giustiniano Martinioni on p. 375 of his expanded edition of Francesco Sansovino's guide to Venice. See Lino Moretti, ed., *Venetia, città nobilissima et singolare* (Venice: Filippi, 1968) for a facsimile reprint of the 1663 edition.

21. Shoduar's numismatic collection is described in Domenico Sestini, *Descrizione d'alcune medaglie greche del museo del Signore Barone Stanislao di Chaudoir* (Florence: Guglielmo Piatti,

1831), vii. Sestini says that the collection had been put together in the period 1816–31 and that Shoduar had made his purchases in Italy, France, Germany, and Crimea. It is likely that Shoduar also purchased the Sagredo portrait on one of these trips. See the introduction to volume 2 of his *Aperçu sur les monnaies russes et sur les monnaies étrangères qui ont eu cours en Russie: Depuis les temps les plus reculés jusqu'à nos jours* (St. Petersburg: F. Bellizard. 1836–37), 3 vols.

22. There are also various cataloguing numbers on the portrait, but none helps us understand its whereabouts before the Shoduar purchase.

23. A445; oil on canvas, 113.9 × 101 cm; WA1935.97. Christopher Lloyd, *A Catalogue of the Earlier Italian Paintings in the Ashmolean Museum* (Oxford: Clarendon Press, 1977), 28–29. The purchase of the painting, described as "a magnificent portrait of a Venetian senator by Leandro Bassano," is reported in *Apollo* 22 (1935): 179.

24. Arslan, *I Bassano*, 1:243. The identification of the sitter as Paolo Nani is stated in Bernard Berenson, *Italian Pictures of the Renaissance; A list of the Principal Artists and their Works, with an Index of Places. Venetian School* (London, Phaidon Press, 1957), 23, apparently on the authority of the artist and print historian Fabio Mauroner.

25. Arslan, *I Bassano*, 1:243.

26. Patricia Fortini Brown, *Private Lives in Renaissance Venice: Art, Architecture, and the Family* (New Haven: Yale University Press, 2004), 11. David Chambers, "Merit and Money: The Procurators of St. Mark and their Commissioni 1443–1605," *Journal of the Warburg and Courtauld Institutes* 60 (1997): 26. See also Giovanni Grevembroch, *Gli abiti de veneziani di quasi ogni età con diligenza raccolti e dipinti nel secolo XVIII*, introd. Giovanni Mariacher (Venice: Filippi, 1981), vol. 1, plate 31, for an image of a senator, and plate 39 for a procurator of St. Mark. Cesare Vecellio reproduces a different robe worn specifically by the consul in Syria, but only as an example of outmoded diplomatic attire: *Habiti antichi, et moderni di tutto il mondo* (Venice: Sessa, [1598]).

27. Carlo Ridolfi's influential biographical sketches of Venetian painters, published in 1648, provide us with a near-contemporary account of the Bassano *bottega*; one of the many private portraits by Leandro he mentions is "a portrait of Nicolò Sagredo in Consular dress." While it is possible that Leandro painted Gianfrancesco's father Nicolò, who had been *provveditore* in Crete and Palmanova and a procurator of St. Mark, but never a consul, it seems more likely that Ridolfi confused Gianfrancesco for his father, and Leandro for his brother. Ridolfi also says that Leandro "produced several works for him," and while we know that this is true for Gianfrancesco, there is no record of Nicolò commissioning or even knowing Leandro. Carlo Ridolfi, "Vita di Leandro da Ponte da Bassano Pittore e Cavaliere," 165–71 in *Le meraviglie dell'Arte ovvero le vite degli illustri pittori veneti e dello Stato*, edited by Detlev von Hadeln (Rome: Società multigrafica editrice SOMU, 1965), 168.

28. On the *Commissioni Dogali*, see Helena Szépe, "Civic and Artistic Identity in Illuminated Venetian Documents," *Bulletin du Musée hongrois des beaux-arts* 95 (2002): 59–78; Helena Szépe, "Distinguished among Equals: Repetition and Innovation in Venetian Commissioni," in *Manuscripts in Transition: Recycling Manuscripts, Texts and Images*, edited by Brigitte Dekeyzer and Jan Van der Stock (Leuven: Peeters, 2005), 441–47; and Laura Nuvoloni, "Commissioni Dogali: Venetian Bookbindings in the British Library," in *For the Love of the Binding: Studies in Bookbinding History Presented to Mirjam Foot*, edited by David Pearson (London: British Library & Oak Knoll Press, 2000), 81–109. I would like to thank Helena Szépe for confirming that the book is a *Commissione Dogale*, and also for kindly checking her database of extant *commissioni* for Gianfrancesco Sagredo.

29. The kilim in the Ashmolean portrait is discussed in John Mills, "The Coming of the

Carpet in the West," in *The Eastern Carpet in the Western World from the 15th to the 17th Century*, ed. Donald King and David Sylvester (London: Arts Council, 1983), 17.

30. Sigismund later features in Galileian iconography, connected to the early trumpet-shaped telescopes. See Marvin Bolt and Michael Korey, "Trumpeting the Tube: A Survey of Early Trumpet-Shaped Telescopes," in *Der Meister und die Fernrohre: Das Wechselspiel zwischen Astronomie und Optik in der Geschichte*, ed. Jürgen Hamel, Rolf Riekher, and Inge Keil (Frankfurt: Harri Deutsch Verlag, 2007).

31. On these tapestries, see Tadeuz Mankowski, "Some Documents from Polish Sources Relating to Carpet Making in the time of Shah Abbas," in *A Survey of Persian Art from Prehistoric Times to the Present*, ed. Arthur Upham Pope (London: Oxford University Press, 1939), 2431–36. The Munich kilim is illustrated in the same work, plate 1265. For a general introduction to the kilims, see Jon Thompson, "Early Safavid Carpets and Textiles," in *Hunt for Paradise: Court Arts of Safavid Iran, 1501–1576*, ed. Jon Thompson and Sheila R. Canby (London: Thames & Hudson, 2003).

32. A letter from Abbas to Sagredo is printed in Favaro, *Giovanfrancesco Sagredo*, 91–92; Sagredo promised an analysis of the shah's dramatic renewal of Persia in his "Relazione" of 4 July 1611 (ibid., 101). He refers to his correspondence with Abbas in a letter to Galileo from Aleppo dated 28 October 1609 (*OG* 10:262).

33. *OG* 11:379.

34. Sagredo's request for a dog from Galileo in 1619 mentions his "rugs and tapestries" (*OG* 12:456), though this would be standard interior decoration for a Venetian palace.

35. "3077 Ritratto d'incognito—Tintoretto—Palazzo Doria—Appartamento privato—Anderson" (Anderson collection, now part of the Alinari archives in Florence). James Anderson (1813–77) and his son Domenico (1854–1938) photographed much of Italy, especially its Roman ruins, and also compiled semi-systematic records of private painting collections. In the 1890s a few hundred paintings from the Doria Pamphilij collection were photographed, and between a Ribera and a Cimabue (in the archive's catalog, if not in the storeroom or gallery) sat a third version of Sagredo, by then attributed to Tintoretto. Some of the paintings carry captions, but they do not seem to have been selected for either an auction catalog (as most are still in the collection) or a visitors' guide (as some were in the private apartments). Only three paintings from the Anderson Archive identify their location as the private apartments, rather than the collection or gallery, of the Palazzo Doria Pamphilij: the Sagredo portrait and two unidentified groups of muses. Many thanks to John Stuart Mills for bringing this image to my attention.

36. The tapered obelisk, resembling Carpaccio's famous adaptations of the Holy Land woodcuts in Bernhard von Breydenbach's *Peregrinatio in terram sanctam* (Mainz Erhard Reuwich, 1486), is also visible in Leandro Bassano's *Carnival* (Vienna), reproduced in Gerola, *Bassano*, 127. It is possible that the obelisk is intended to represent the column of Diocletian, known as "Pompey's Pillar," and that the rotunda is meant to recall Alexandria's Serapeum.

37. See Deborah Howard, *Venice & the East: The Impact of the Islamic World on Venetian Architecture 1100–1500* (New Haven: Yale University Press, 2000), 76 and 94. The classic studies are Hermann Thiersch, *Pharos, Antike, Islam und Occident: Ein Beitrag zur Architekturgeschichte* (Leipzig and Berlin: B. G. Teubner, 1909) and Peter Marshall Fraser, *Ptolemaic Alexandria* (Oxford: Clarendon Press, 1972). For a contemporary account of Alexandria, see Mikołaj Krzysztof Radziwiłł's *Hierosolymitana Peregrinatio* (Brunswick: 1601), 203–10.

38. Printed in Latin in 1596 (Venice: Apud haeredes Simonis Galignani de Karera) and in Italian in 1597–98 (Venice: Appresso Gio. Battista & Giorgio Galigani, fratelli).

39. *Geografia* (Venice: 1596), v.2, p. 152.

40. See *Galileo's Glassworks* (Cambridge, MA: Harvard University Press, 2008), 38–46 and 72–78. For an invigorating essay on the importance, and loss, of Alexandrian science, see Lucio Russo, *The Forgotten Revolution: How Science Was Born in 300 B.C. and Why It Had to Be Reborn* (Berlin: Springer, 2004), original edition 1996, as well as Fraser's *Ptolemaic Alexandria.*

41. Giovan Battista Della Porta, *Magiae naturalis libri XX* (Naples: Apud Horatium Saluianum, 1589), 270.

42. In Sagredo's description of his plan to have Leandro Bassano paint a Sagredo family portrait, he says, echoing a lost comment by Galileo, that Bassano is an excellent portraitist, but "in the invention and gestures somewhat rustic," and he asks Galileo to look out for a Florentine painter of "spirit and ingenuity" to help with the composition. Sagredo provides the general content of the picture: a Madonna, Saint Gerardo Sagredo, Zaccaria, his wife, their six boys and one girl, and also five dead boys and one girl who are to be depicted as little angels. He also stipulates the scale (life size) and the canvas size. With these details, Galileo is to find an artist to compose a preliminary sketch—a job he actually took on, sending two drafts in the summer of 1619. See *OG* 12:454 and 461.

43. On Leandro's melancholia, exhibitionism, and paranoia, see Ridolfi, *Vita*, 170.

44. *OG* 12:415–16. Arno died of rabies in March 1619; *OG* 12:446.

45. *OG* 12:418.

46. *OG* 12:419. The *DBI* entry on Leandro describes his "portrait monopoly" in Venice in the first two decades of the seventeenth century. For a partial list of Leandro's portrait sitters, see Ridolfi, *Vita*, 170. For hints of Leandro's contacts with Paduan intellectuals, see Raffaele Dondi, "Di Leone Bonzio incisor pubblico di anatomia a Venezia e del suo ritratto dipinto da Leandro da Ponte detto il Cavalier Bassano," *Rivista di Storia della Medicina* 4 (1960). Leandro also painted Galileo's colleague Prospero Alpini. Critical dismissal of Gerolamo's talents started early. Ridolfi, *Vita*, 171, was probably the first to depict him as merely a good copier of his father's and brothers' work, though Jacopo's will hints at this view. The documentation used in the present chapter indicates that a reexamination of Gerolamo's oeuvre is overdue. None of Gerolamo's extant works is dated.

47. *OG* 12:448.

48. *OG* 12:452.

49. *OG* 12:454.

50. *OG* 12:460.

51. *OG* 12:464–5.

52. "52 Un ritratto de un Sagredo finito dredo detta Madonna del quondam signor Geronimo" (p. 50), "408 Un retratto del signor Zuan Francesco Sagredo finito de man del signor Geronimo" (p. 62), and "226–227 Il Sagredo et Apolo" (p. 84) in Stefania Mason, ed., "L'Inventario di Gerolamo Bassano e l'Eredità della Bottega." Special issue, *Notiziario dell'Associazione "Amici dei Musei e dei Monumenti di Bassano del Grappa"* (2009).

53. See, for example Sergio Pagano, *I documenti vaticani del processo di Galileo Galilei*, 2nd edn. (Vatican City: Archivio Segreto Vaticano, 2009); John Heilbron, *Galileo* (Oxford: Oxford University Press, 2010); David Wootton, *Galileo: Watcher of the Skies* (New Haven: Yale University Press, 2010). Art historians have been less accepting: Horst Bredekamp's *Galilei der Künstler: Der Mond. Die Sonne. Die Hand* (Berlin: Akademie Verlag, 2007) remains uncommitted, without explanation. More interesting is Meri Sclosa, "Vedute possibili: Finestre paesaggistiche nella ritrattistica di Leandro Bassano e Domenico Tintoretto," *Paragone* 94 (2010), who proposes instead that the sitter is in fact a different Giovanni Francesco Sagredo (son of Piero), who was made Duke of Crete in 1602. Chania is offered as the background view in the Ashmolean por-

trait. But Sclosa misrepresents my argument that the window scene depicts ancient Alexandria by showing how it does not depict seventeenth-century Alexandretta, which was precisely my point: it depicts ancient Alexandria. Sclosa's counter-evidence that it represents contemporary Chania is entirely unconvincing: Chania's domed tower on a rocky pier might be any lighthouse, whereas the Ashmolean portrait clearly shows the square jetty of the Alexandrian Pharos. Chania, as Sclosa's other illustrations make perfectly clear, was a small and confined maritime fortress (constructed by Sanmicheli in the 1560s), with several easily identifiable features: the famous arsenals, the San Salvatore bastions, and the lighthouse on the port side for entering vessels. None of these elements is visible in the Ashmolean painting. The city depicted there, rather, is unwalled and extends towards the surrounding hills; the lighthouse is on the starboard side, as it was in Alexandria. Moreover, as Monique O'Connell's *Men of Empire* demonstrates, several branches of the Sagredo family had a long history of positions on Crete, but their tie was to Heraklion, not Chania. The other Giovanni Francesco Sagredo (son of Piero) built a fountain in Heraklion, which is still extant. Why, then, would Leandro Bassano choose to paint an unrecognizable Chania?

54. Paris, Bibliothèque nationale, MS Latin 11195, f. 57, published in Karolina Targosz, " 'Le Dragon Volant' de Tito Livio Burattini," *Annali dell'Istituto e Museo di storia della scienza di Firenze* 2 (1977): 72. See also René Taton, "Nouveau document sur le 'Dragon Volant' de Burattini," *Annali dell'Istituto e Museo di storia della scienza di Firenze* 7 (1982).

55. The source of the story is uncertain, but it probably comes from a vague anecdote of an anonymous airborne Venetian launching from the *campanile* of St. Mark's, reported in Johann Sturm, *Linguae latinae resolvendae ratio* (Strassburg: 1581), 40, via Friedrich Hermann Flayder's 1627 *De Arte Volandi* (Tubingen: 1627), 47. John Wilkins also reports the story in *Mathematicall magick* (London: 1648), 3, citing Joann Ernest Burggrav, *Achilles πανοπλος Redivivus, seu Panoplia Physico-Vulcania quâ in praelio φιλοπλος in Hostem educitur Sacer et inviolabilis* ([Amsterdam]: 1612). See Natalie Kaoukji, "Flying to Nowhere: Mathematical Magic and the Machine in the Library" (PhD dissertation, Cambridge University, 2008).

56. Sagredo's anti-Jesuitism is often mistaken for atheism. Zaccaria sent Sarpi a note saying he would send a boat to bring him to Gianfrancesco. ASV, *Consultori in iure*, 25, c. 9 (145v–146r), first printed in Bartolomeo Cecchetti, *La republica di Venezia e la corte di Roma nei rapporti della religione* (Venice: Naratovich, 1874) 2: 448. Many thanks to Corrado Pin for this reference. Sarpi himself suffered a similar fate, becoming an illustrious atheist in the hands of his best modern anglophone biographer, David Wootton, *Paolo Sarpi: Between Renaissance and Enlightenment* (Cambridge: Cambridge University Press, 2002). Sarpi's death was carefully witnessed for evidence of irreligion: see Gaetano Cozzi "Sulla morte di fra Paolo Sarpi," in *Miscellanea in onore di Roberto Cessi*, II (Rome, Edizioni di storia e letteratura, 1958), 387–96.

57. Sagredo's will is printed in Favaro, *Giovanfrancesco Sagredo*, 272–76.

58. *OG* 13:42.

59. *OG* 13:49.

60. *OG* 13:45. The list of thirty-three paintings includes subjects such as "cheese," "a plate of oysters," "artichokes," and "a goose," as well as five landscapes.

61. Favaro claimed at the end of the nineteenth century that the Sagredo collection had only recently been dispersed, but I can find no source for this statement. Archival records give us some hint of a collection's survival into the eighteenth century. See, for example, the intriguing apparent coherence, perhaps illusory, of the following documents, located in an eighteenth-century inventory of a Sagredo library: "n.192 Le Statue Parlanti, con altri cose insieme unite curiose. In fol.; n. 194 Aforismi, e scritture di F. Tomasso Campanella Consultor della Repub.

in fol.; n. 196 Trattato della Seda Nostrana di Gio.Batt.a Follo. in fol.; n. 197 Descrizione delle Provincie di Francia, Trattato d'Astrologia in Dialogo. fol.; n. 198 Dispacci Publici de'Residenti in Alessandretta, d'Aleppo, ed altri luoghi del 1608, 1609, 1610, 1611 in fol." The last object almost certainly derives from Gianfrancesco; the others refract many of his interests. One of the manuscripts, the "Trattato della Seda Nostrana di Gio.Batt.a Follo. in fol.," is potentially traceable; surviving contenders are at Princeton (Ms. C0199), "Trattato della seda nostrana di Gio[vanni] Batt[ista] Follo"; Herzog August Bibliothek, Wolfenbuttel (Cod. Guelf. 28.6 Aug. 4°), "Sommario del trattato della regolatione del dado delle sede nostrane conforme al raccordo proposto nell eccellentissimo senato da Giovanni Battista Follo (Bombardello) Vicentino," bound with "Riposta al trattato et ragioni di Giov. Battista Follo sopra il raccordo di accrescer il dacio della seda dell illustrissimo Signore Agostino dal Bene" (del Bene was a Venetian consultor and supporter of Donà during the Interdict), and Correr (Ms. Donà delle Rosa, 217), "Esposizione del Raccordo di Giovan Battista Follo cittadino veneciano intorno alle sede." The tract contains a dialogue between Sebastiano Venier, one of Sagredo's closest friends; Francesco Mocenigo, Bruno's final patron; the artist Giacomo Galli; and the lawyer Labieno Vellutello. Other inventories also give clues to the afterlife of Gianfrancesco's papers: Correr, Mss. P.D. C.2193, V, an inventory from 1760 ("Inventario delle Carte attinenti alla Facoltà Sagredo, e Grimani Calergi Disposte per ordine" contains, at n. 1603, "Anno 1608 Quadernetto N.H.S. Francesco Sagredo."

62. Zaccaria's efforts were frustrated; his son Nicolò became interested in Copernicanism in the 1630s, to Galileo's great joy. See *OG* 16:411 and 431.

Chapter Two

1. Antonio Favaro, in a rare act of editorial bias, declined to publish Galileo's horoscopes in full in his magisterial edition of Galileo's complete works, but noted that there were several references within Galileo's surviving account books for payment for this service in 1603 from Paduan students; one of these refers to an extant horoscope archived alongside Sagredo's. *OG* 19:205–6.

2. Antonio Favaro, ed., *Carteggio Inedito di Ticone Brahe, Giovanni Keplero e di altri celebri astronomi e matematici dei secoli XVI e XVII con Giovanni Magini* (Bologna: Zanichelli, 1886), 309 (3 June 1605, Strassoldo to Magini).

3. BNCF mss. Gal. 81, f.13, translation by Deborah Houlding and reproduction of the horoscope at http://www.skyscript.co.uk/galchart2.html. See Julianne Evans, "On the Character of Sagredo: An English translation of Galileo's judgements upon his nativity of Giovanni Sagredo," *Culture and Cosmos* 7 (2003).

4. See, in addition to the older studies of the Venetian aristocracy, Dorit Raines, *L'Invention du mythe aristocratique: L'image de soi du patriciat vénetien au temps de la Sérénissime*, 2 vols. (Venice: Istituto Veneto di Scienze, Lettere ed Arti, 2006); and Monique O'Connell, *Men of Empire: Power and Negotiation in Venice's Maritime State* (Baltimore: Johns Hopkins University Press, 2009).

5. See Antonio Favaro, "Lo Studio di Padova e la Compagnia di Gesù sul finire del secelo decimosesto," *Atti del Reala Istituto Veneto di Scienze, Lettere ed Arti* 4 (1877–78), with a useful documentary appendix, continued by "Nuovi documenti sulla vertenza tra lo Studio di Padova e la compagnia di Gesù sul finire del secolo decimosesto," *Nuovo Archivio Veneto* 21 (1911); Favaro, *Galileo Galilei e lo Studio di Padova*, chapter 3; Gaetano Cozzi, *Paolo Sarpi tra Venezia*

e l'Europa, 144–53; John Patrick Donnelly, "The Jesuit College at Padua: Growth, Suppression, Attempts at Restoration, 1552–1606," *Archivum Historicum Societatis Iesu* 51 (1982); Maurizio Sangalli, "Apologie dei Padri Gesuiti contro Cesare Cremonini" in *Università, accademie, gesuiti: Cultura e religione a Padova tra cinque e seicento* (Padua: Edizioni Lint, 2001), originally published in *Atti e memorie dell'accademia galileiana di scienze lettere ed arti, parte III: Memorie della classe di scienze morali, lettere ed arti* 110 (1997–98): 241–355; Cesare Cremonini, *Le Orazioni*, ed. Antonio Poppi (Padua: Antenore, 1998); Paul Grendler, *The Universities of the Italian Renaissance*, (Baltimore: Johns Hopkins University Press, 2004), 479–83; Edward Muir, *The Culture Wars of the Late Renaissance: Skeptics, Libertines, and Opera* (Cambridge MA: Harvard University Press, 2007); the first full account forms the climax of book four of Antonio Riccobono's *De Gymnasio Patavino . . . Commentariorum Libri Sex* (Padua, 1598; facsimile reprint Bologna: Arnaldo Forni, 1980) 103r–106v.

6. See Aldo Stella, "Galileo, il circolo culturale di Gian Vincenzo Pinelli e la 'Patavina Libertas'" in *Galileo e la Cultura Padovana*, ed. Giovanni Santinello (Padua: CEDAM, 1992), 307–25, cited in Michele Camerota, *Galileo Galilei e la cultura scientifica nell'età della controriforma* (Rome: Salerno, 2004), 76–77.

7. Differing evaluations of the impact of Padua's intellectuals on the surrounding area in this period is at the heart of Paola Zambella's harsh critique of Carlo Ginzburg's *The Cheese and the Worms*: "Uno, due, tre mille Menocchio?" *Archivio storico italiano* 137 (1979): 51–90. Ginzburg responded to the charges in the first English edition (1992), 154–56.

8. Paolo Gualdo, *Vita Ioannis Vincentii Pinelli, patricii genuensis* (Augsburg: 1607), 71, cited in Camerota, *Galileo Galilei*, 77.

9. *OG* 9:414.

10. On Barozzi, see Paul Rose, "A Venetian Patron and Mathematician of the Sixteenth Century: Francesco Barozzi (1537–1604)," *Studi Veneziani* 1 (1977): 119–80, and Ugo Baldini and Leen Spruit, *Catholic Church and Modern Science: Documents from the Archives of the Roman Congregations of the Holy Office and the Index* 4 vols. (Rome: Libreria Editrice Vaticana, 2009), 1:597; on Cremonini, see Antonino Poppi, *Cremonini, Galilei, e gli inquisitori del Santo a Padova* (Padua: Centro Studi Antoniani, 1993); Leen Spruit, "Cremonini nelle carte del Sant'Uffizio romano" in *Cesare Cremonini. Aspetti del pensiero e scritti*, ed. Ezio Riondato and Antonino Poppi, 2 vols. (Padua: Accademia Galileiana di Scienze, Lettere e Arti, 2000).

11. Reproduced in Favaro, *Lo Studio di Padova e la Compagnia di Gesù*, Doc. V.

12. Ibid.

13. Muir, *The Culture Wars*, 25.

14. Favaro, *Galileo Galilei e lo studio di Padova*, 68.

15. Favaro, "Lo Studio di Padova e la Compagnia di Gesù," 495.

16. Ibid., 491.

17. Cesare Cremonini, "*Lecturae Exordium*" in *Le Orazioni*, ed. Antonio Poppi, 3–51 (Padua: Antenore, 1998), 44.

18. See, for example, the discussion of Cremonini's speech in Riccobono's *De Gymnasio Patavino . . . Commentariorum Libri Sex*, 103v.

19. William J. Bouwsma, *Venice and the Defense of Republican Liberty: Renaissance Values in the Age of the Counter Reformation* (Berkeley and Los Angeles: University of California Press, 1968), notes in the introduction the effect on the writing of the book of the "disorders that have agitated Berkeley since 1964"; a decade later, in his AHA Presidential Addresses, "The Renaissance and the Drama of Western History," *American Historical Review* 84 (1979): 1, Bouwsma

reevaluated the late Sixties' impact on "Renaissance" historiography. Bouwsma was hardly a radical; the history of the development of microhistory within the context of 1960s Italian and, to a lesser extent, American politics, especially analyzing the crucial contributions of Edward Muir and Guido Ruggiero, would be well worth telling.

20. Sangalli, *Università, accademie, gesuiti*, 80–81, 96, 130, and 153.

21. Sangalli claims that the only extant copies of the Jesuit texts are in ARSI, Ven. 105. Donnelley, "The Jesuit College at Padua," 46 and 63, uses the same volume but notes the presence of copies of four of the refutations (by Ludovico Gagliardi, Benedetto Palmio, Giovanni Domenico Bonaccorsi, and Paolo Comitoli [under the pseudonym Eufemio Filarete]; that of Antonio Possevino is lacking), without the names of their authors, in Ambrosiana D. 463 Inferior. I have not been able to consult this collection, though presumably the refutations were collected by Pinelli.

22. On the identification of the young priest as De Dominis, future scientist, polemicist, and double apostate, see Donnelley, "The Jesuit College at Padua," 54n45. De Dominis responded to the theoretical inadequacies of the *Sidereus nuncius*'s treatment of optics with the *De radiis visus et lucis in vitris perspectivis et iride tractatus* (Venice: Baglioni, 1611) and to the failure of Galileo's (manuscript) tidal argument, the *Flusso e reflusso del mare* (1616) with *Euripus, seu De fluxu et refluxu maris sententia* (Rome: 1624).

23. See, especially, Harold Love, *Scribal Publication in Seventeenth-Century England* (Oxford: Oxford University Press, 1993), H. R. Woudhysen, *Sir Philip Sidney and the Circulation of Manuscripts, 1558–1640* (Oxford: Oxford University Press, 1996) and Brian Richardson, *Manuscript Culture in Renaissance Italy*.

24. Filippo De Vivo, *Information and Communication in Venice: Rethinking Early Modern Politics* (Oxford: Oxford University Press, 2007), and Filippo De Vivo, *Patrizi, informatori, barbieri: Politica e comunicazione a Venezia* (Milan: Feltrinelli, 2012), with a full bibliography of Interdict publications.

25. Donnelley, "The Jesuit College at Padua," 71–77.

26. ARSI, Ven. 5 Epp. P.is. Gen. f.2 (13/8/1600)

27. ARSI, Ven. 5 Epp. P.is. Gen. f.8 (29/01/1600/01).

28. ARSI, Ven. 5 Epp. P.is. Gen. f.125 (14/10/1601)

29. *DBI* enty. Gagliardi died on 6 July 1607. In July 1606, according to the English ambassador to Venice, Sir Henry Wotton, a price had been set on his head as a rebel for preaching against Venice in Mantua. Logan Pearsall Smith, ed. *The Life and Letters of Sir Henry Wotton*, 2 vols. (Oxford: Clarendon Press, 1907), 1:355.

30. ARSI, Ven. 5 Epp. P.is. Gen. f.132r, 17 November 1601. Similar letters, with the same date, are addressed to Giacomo Alvise Cornaro (f.132r), Lodovico and Giovanni Andrea Basso (ff.132rv) and Gianpaolo Contarini (f.132v). Paolo Valle (Vallius) is well known to Galileianists as one of the sources of Galileo's notes on logic, taken either directly from manuscript sources while at Pisa (according to William Wallace in "The Dating and Significance of Galileo's Pisan Manuscripts" in *Nature, Experiment, and the Sciences: Essays on Galileo and the History of Science in Honour of Stillman Drake*, ed. Trevor H. Levere and William R. Shea [Dordrecht: Kluwer, 1990], and later publications) or from a plagiarized print source after 1597 (according to Adriano Carugo and Alistair C. Crombie, "The Jesuits and Galileo's Ideas of Science and of Nature," *Nuncius* 8 (1983). For a good summary of the disagreement, see Antonio Nardi's review of Michele Camerota's *Gli Scritti "De Motu Antiquiora" di Galileo Galilei: Il Ms Gal 71: Un'analisi storico-critica* in *Nuncius* 10.2 (1995): 808–12.

31. eg. ARSI, Ep. Gen. Ven. S II, 297v, 312v.

32. *OG* 2:536; *OG* 19:127 and 194.

33. John O'Malley, *The First Jesuits* (Cambridge, MA: Harvard University Press, 1993).

34. Indeed, there is much evidence from the 1590s to suggest that heterodoxy was tolerated within the Society of Jesus, before the clampdown in the early seventeenth-century. See Ugo Baldini, *Legem impone subactis: Studi su filosofia e scienza dei Gesuiti in Italia, 1540–1632* (Rome: Bulzoni, 1992).

35. See the excellent introduction by Marcella Grendler, "Book Collecting in Counter-Reformation Italy: The Library of Gian Vincenzo Pinelli (1535–1601)," *The Journal of Library History* 16 (1981).

36. See Leonardo Garzoni, *Trattati della calamita*, ed. Monica Ugaglia, (Milan: Franco Angeli, 2005) and Monica Ugaglia "The Science of Magnetism before Gilbert: Leonardo Garzoni's Treatise on the Loadstone," *Annals of Science* 63 (2006).

37. Sarpi's natural philosophical jottings, especially those known as the "Pensieri," have been forcibly transferred to this flimsy context.

38. Stella, "Galileo, il circolo culturale di Gian Vicenzo Pinelli," 307–25.

39. On Tycho's epistolary networks, see Adam Moseley, *Bearing the Heavens: Tycho Brahe and the Astronomical Community of the Late Sixteenth Century* (Cambridge: Cambridge University Press, 2007).

40. On this episode, see Wilhelm Norlind, "Tycho-Brahe et ses rapports avec l'Italie," *Scientia: Rivista di scienzia* 49 (1955); and Massimo Bucciantini, *Galileo e Keplero: Filosofia, cosmologia e teologia nell'Età della Controriforma* (Turin: Einaudi, 2003). For Tengnagel, see also Josef Smolka, "The Scientific Revolution in Bohemia," in *The Scientific Revolution in National Context*, ed. Roy Porter and Mikulas Teich (Cambridge: Cambridge University Press, 1992), 210–39.

41. 3 January 1600, Brahe to Pinelli in *Tychonis Brahe dani opera omnia*, ed. Johan Ludwig Emil Dreyer and Hans Raeder, vol. 3 *Epistolae astronomicae* (Amsterdam: Swets and Zeitlinger, 1972), 228.

42. Favaro, *Carteggio inedito*, 258–59, Tengnagel to Magini (1603?). The implied reference to Sagredo has generally been accepted, though there is a possibility that Sarpi was intended.

43. On this entire episode, see the excellent account in Bucciantini, *Galileo e Keplero*, 83–91 and 117–119.

44. See the introduction (ix–xxxviii) to Simon Schaffer et al., eds. *The Brokered World: Go-Betweens and Global Intelligence, 1770–1820* (Sagamore Beach, MA: Science History Publications, 2009). For an interesting case study including the Barberini and Galileo, see Janie Cole, "Cultural Clientelism and Brokerage Networks in Early Modern Florence and Rome: New Correspondence between the Barberini and Michelangelo Buonarroti the Younger," *Renaissance Quarterly* 60 (2007).

45. Contarini would be among the group of senators to accompany Galileo to the viewing platform of the campanile of San Marco for his telescope demonstration a decade later.

46. *OG* 10:77.

47. Magini dedicated maps of Trevigiano and Bergamo to Sagredo in his posthumously published *Italia* (1620).

48. Horatio F. Brown, ed. *Calendar of State Papers Relating to English Affairs in the Archives of Venice, Vol. 9, 1592–1603* (London: Eyre and Spottiswoode, 1897), henceforth *CSP*, 514, doc. 1113.

49. *OG* 10:101.

50. Roger Gaskell is currently working on a reconstruction of the contents of the pre-fire library.

51. William Barlow, *Magnetical Advertisements Concerning the Nature and Property of the Loadstone* (London: 1616), 87–88 (no pagination). No archive of Barlow's own letters seems to have survived. See Anita McConnell's entry in *ODNB*.

52. Valleriani's *Galileo Engineer*, a fine study of mechanical socioepistemology around the Arsenal, which deals in depth with Sagredo's contributions to Galileo's optics, thermoscopy, and mechanics, neglects their joint study of magnetism. This reflects a more general historiographical tendency to either place magnetism solely in the field of natural magic or deal with it merely as a precursor to late-nineteenth-century electromagnetic studies.

53. Simon Schaffer, "Newton on the Beach: The Information Order of the *Principia Mathematica*," *History of Science* 47 (2009): 246.

54. *CSP*, 9: 514–16 (docs. 1113–15), 526–534 (docs. 1131–35), 564–69 (docs. 1169–71).

55. See Garzoni, *Trattati della calamita*, and my review in *History of Universities* 13 (2008): 221–23.

56. *OG* 10:91. In 1608 Sarpi wrote to Groslot that he thought only two people had written with any originality in his time: François Viète and William Gilbert. Paolo Sarpi, *Lettere ai Protestanti*, ed. Manlio Duilio Busnelli, 2 vols. (Bari: Laterza, 1931), 1:16–17.

57. The implications of magnetic declination are more complex than usually presumed; a letter from Kepler to an unidentified friend of Galileo (Pinelli or Magini?), written in July 1599, hints that accurate measurements of declination might be used to test the claim made by Copernicus's teacher Domenico Maria Novara that the terrestrial (rather than magnetic) poles were gradually shifting (*OG* 10:75). This was a heavily disputed claim, reported in Magini's *Tabulae secundorum mobilium coelestium* (Venice: Damiani Zenarii, 1585), 29, and rejected in Gilbert's *De Magnete*, 213. See Favaro, *Carteggio inedito*, 80–82. Gilbert's rejection of Novara was of interest to Galileo; his copy of Gilbert contains a single approving note, transcribed in *OG* 8:625. Doubts of the permanent accuracy of classical astronomical data threatened the foundations of the early modern venture; on Tycho Brahe's attempt to send a mission to Alexandria to check Ptolemy's readings, see above.

58. *OG* 10:101.

59. For these two readings of Gilbert, see Edgar Zilsel's classic essay "The Origins of William Gilbert's Scientific Method," *Journal of the History of Ideas* 2 (1941); and the excellent response by John Henry, "Animism and Empiricism: Copernican Physics and the Origins of William Gilbert's Experimental Method," *Journal of the History of Ideas* 62 (2001).

60. Paolo Alberi Auber, "Una miniera, un forno per il ferro e due uomini di Scienza fra le montagne: Nicola Cusano e Gianfrancesco Sagredo," *Archivio per l'Alto Adige* 100 (2006): 30, citing an unspecified document from the Registri del Comune di Vodo di Cadore. Auber's work, while providing many new archival sources, should be used with caution.

61. Alberi Auber, "Una miniera, un forno," 37, citing figures from an otherwise unspecified document referred to as the *libro mastro della miniera* in the Archivio Vescovile di Bressanone. For 1603 there is an annual amount of 3740.5 kübel, equivalent to 1246 fiorini and 50 kreuzer, to Zaccaria Sagredo.

62. Sagredo's will, dated 14 July 1608, is at ASV, Notarile, Testamenti, b. 344n372, and is printed in Favaro (1902). There is another copy, from the eighteenth century, at Correr, Mss. P.D. C.2192, II/3.

63. See Domenico Bertoloni Meli, *Thinking with Objects: The Transformation of Mechanics in the Seventeenth Century* (Baltimore: Johns Hopkins University Press, 2006), chapter 1.

64. For an accessible account of the consolidation and transformation of magnetic philosophy, see Stephen Pumfrey, *Latitude and the Magnetic Earth: The True Story of Queen Elizabeth's Most Distinguished Man of Science* (Cambridge: Icon Books, 2002).

Chapter Three

1. Heilbron, *Galileo*; W. R. Laird, "Archimedes among the Humanists," *Isis* 82 (1991); Paul Rose, *The Italian Renaissance of Mathematics: Studies on Humanists and Mathematicians from Petrarch to Galileo* (Geneva: Droz, 1975); Mark A. Peterson, *Galileo's Muse: Renaissance Mathematics and the Arts* (Cambridge, MA: Harvard University Press, 2011), chapter 1.

2. See Stillman Drake, "Tartaglia's Squadra and Galileo's Compasso," *Annali dell'Istituto e Museo di Storia della Scienza di Firenze* 2 (1977), as well as his introduction to *Operations of the Geometric and Military Compass, 1606* (Washington: Dibner Library of the History of Science and Technology, 1978); and Filippo Camerota, *Il Compasso di Fabrizio Mordente: Per la storia del compasso di proporzione* (Florence: Olschki, 2000)

3. On scribal publication in the period, see Richardson, *Manuscript Culture*, and his special issue of *Italian Studies*, 66 (2011), on "The Uses of Manuscripts in Early Modern Italy."

4. Alarico Carli and Antonio Favaro, eds., *Bibliografia galileiana 1565–1895* (Rome: [s.n.], 1896), include an entry, n. 19, for a reprint of the work in the same year in Verona by Bartolomeo Merlo.

5. *Discorso dell'Ecc. sig. Antonio Lorenzini da Montepulciano intorno alla nuova stella* (Padova: Pasquati, 1605), reprinted in *OG* 2.

6. Galileo makes the claim in his *Difesa, OG* 2:521.

7. Padua: 1605. Reprinted, with Galileo's occasionally obscene annotations, in *OG* 2:287–305.

8. *OG* 2:292.

9. Drake, *Operations*, 104–10; Camerota, *Il Compasso*, 117–21.

10. Priority disputes between Galileo and Mayr revived after the publication of the *Sidereus nuncius*. See the recently discovered almanac written in 1611 which contains Mayr's description of his observations of the satellites of Jupiter, described in Klaus-Dieter Herbst, "Galilei's Astronomical Discoveries Using the Telescope and Their Evaluation Found in a Writing-Calendar from 1611," *Astronomische Nachrichten* 330 (2009).

11. *OG* 2:283–84 and 524, where it forms part of the *Difesa*.

12. For a different account of these overlapping economies, see Mario Biagioli, *Galileo's Instruments of Credit* (Chicago: University of Chicago Press, 2006), 7–13.

13. See *OG* 19:131–46 for Galileo's accounts with Mazzoleni from 1599 to 1610.

14. *OG* 19:147. The lower price is calculated on the sale of a compass together with a "compasso da quattro punte" for a total of forty-two lire. The standard price of the latter instrument was eight lire.

15. *OG* 19:149–58.

16. Unfortunately, the accounts with his unruly copyist, Silvestro, are fragmentary and unclear, and it is impossible to calculate how much he paid for each manuscript. See *OG* 19:166.

17. *OG* 10:149.

18. An illuminating parallel case is Mutio Oddi, whose career has been excellently reconstructed in Alex Marr's *Between Raphael and Galileo: Mutio Oddi and the Mathematical Culture of Late Renaissance Italy* (Chicago: University of Chicago Press, 2011). There may, in fact, be points of direct contact between Sagredo and Oddi; one of Oddi's students, listed as "Luigi (?)

Solaro—unidentified," who took a course of rudimentary mathematics with Oddi in 1617, might be one "Sr. Solari" referred to by Sagredo in a letter to Cardinal Federico Borromeo dated Venice, 10 November 1612. The letter is now in the Biblioteca Ambrosiana, Milan, at Ms. G-253 inf., f.136rv. A scan of the microfilm copy held at Harvard was generously supplied by John Collins at the Lamont Library. Sagredo stopped in Milan on his way back to Venice from Aleppo via Marseilles in 1611.

19. *OG* 2:533.

20. *OG* 19:154.

21. *OG* 19:166.

22. *Le Operazioni del compasso geometrico e militare di Galileo Galilei nobil Fiorentino lettor delle matematiche nello Studio di Padova: Dedicato al Sereniss. Principe di Toscana D. Cosimo Medici* (Padova: In Casa dell'Autore, per Pietro Marinelli, 1606). The Venetian license for the book, dated 26 June 1606, refers to it as the "Division della linea, senza nome dell'autore" (*OG* 19:222–23). It is possible that this is in fact a proposed pirate edition of one of Galileo's manuscript copies.

23. This figure may be false; even taking into account the emergence of several forged copies of the *Operazioni* on the market around 2005, the edition's survival rate (approximately twenty-five copies now extant) implies a larger print run. Galileo's imagined community of readers for the book were non-Latinate soldiers: "Finalmente, essendo mia intenzione di esplicare al presente operazione per lo più attenenti soldato, ho giudicato esser bene scrivere in favella toscana, acciò che, venendo talora il libro in mano di persone più intendenti della milizia che della lingua latina, possa da loro esser comodamente inteso." *OG* 2:371.

24. The first edition to include illustrations was Bernegger's Latin translation of 1612.

25. See *OG* 19:167 for the printing costs of *Le Operazioni*.

26. *OG* 19:158.

27. *OG* 19:146.

28. *OG* 10:348–53.

29. Padua: Apud Petrum Paulum Tozzium, ex typographia Laurentj Pasquati, 1607. Reprinted in *OG* 2.

30. On Capra (1580–1626), see G. Gliozzi's entry, *sub voce*, in the *DBI* and Reeves, *Galileo's Glassworks*, 100–112.

31. *OG* 10:171–72.

32. *OG* 2:533.

33. It is worth giving the full title here: *Difesa di Galileo Galilei Nobile Fiorentino, Lettore delle Matematiche nello Studio di Padova, Contro alle Calunnie & imposture di Baldessar Capra Milanese, Usategli sì nella Considerazione Astronomica sopra la nuova Stella del MDCIIII come (& assai più) nel publicare nuovamente come sua invenzione la fabrica, & gli usi del Compasso Geometrico, & Militare, sotto il titolo di Usus & fabrica Circini cuiusdam proportionis, &c.* (Venice: Presso Tommaso Baglioni, 1607). *OG* 2.

34. *OG* 2:519.

35. *OG* 2:557.

36. Ibid.

37. For good examples of the practice in Renaissance Florence, see Lauro Martines, *April Blood: Florence and the Plot against the Medici* (Oxford: Oxford University Press, 2003).

38. BNCF MD 1028; *OG*, 10: 177 (24 August 1607). Cosimo replied on 11 September 1607, saying he had not yet finished reading the book (*OG* 10:179).

39. The copy resurfaced in 2005, was offered by Peter Harrington of London, and is now in the Library of Congress awaiting cataloguing. For a description, see Jonathan Hill, *Catalogue*, 200n28; Martayan Lan New York Book Fair 2011, catalog 31. Saracinelli received his copy alongside Galileo's letter and copy for Cosimo, and wrote on 11 September saying that he had finished it and already passed it on to a friend, who wanted to show it to someone else (*OG* 10:180–81).

40. Biagioli, *Galileo Courtier*, 24.

41. BNCF, B.R. 165; *OG* 10:181.

42. Oklahoma, History of Science Collections Vault, bound with *Le operazioni del compasso geometrico, et militare*. Amadori seems to have been a close confidant of Galileo's during the writing of the *Sidereus nuncius*: on 18 March 1610, just a few days after the book's publication, Cigoli wrote to Galileo from Rome saying that he had heard from Amadori in Florence about Galileo's observations and his Venetian book. Cigoli read his letter to several Roman acquaintances, including Cardinal Del Monte, who immediately requested both a telescope and a copy of the book (*OG* 10:274). In 1613 Amadori sent Cigoli a list of prohibited books for which he wanted licenses to read. These included the works of Gesner, Fuchs, Cardan, Paracelsus, Boccaccio, and Machiavelli (*OG* 18:415). On the granting of reading licenses, an underused resource for book historians, see the important documents contained in Baldini and Spruit, eds., *Catholic Church and Modern Science*, covering the sixteenth century. Material for the seventeenth century will soon follow.

43. Accademia di Belle Arti, Florence, Archivio Storico, X. IV. 6.

44. Biblioteca universitaria di Padova, S.N. 11910; Antonio Favaro, "Appendice seconda alla Libreria di Galileo," in *Scampoli galileiani* 2:2. On Willoughby, see Antonio Favaro, "Riccardo Willoughby," in *Amici e corrispondenti di Galileo*, ed. Paolo Galluzzi, vol. 2, 1001–5. (Florence: Salimbeni, 1983).

45. University of St. Andrews, Special Collections, QB41.G2D5.

46. Yale University, Beinecke Rare Book and Manuscript Library, QB41 G323 1607.

47. Tomash collection, G5. In addition, the copy at ETH-Bibliothek Zürich, Rar. 4432:3, seems to have lost an inscription; there are no other indications of provenance.

48. This figure is probably too low; it is based almost exclusively on digital catalogues.

49. Galileo to Cosimo de' Medici, 24 August 1607, Padua; *OG* 10:177–78. Cosimo's reply, dated 11 September 1607, Florence, is on 179.

Chapter Four

1. See Edward Muir, *Mad Blood Stirring: Vendetta and Factions in Friuli during the Renaissance* (Baltimore: Johns Hopkins University Press, 1998) for vivid descriptions of the area in the early sixteenth century. The initial plan for Palmanova was drawn up by Giulio Savorgnan, a notable descendent of Muir's vendetta-driven family.

2. ASV, Senato 111, Secreta, Palma 7 (1605–7), 26 October 1605. On Gussoni, see *DBI* entry, *sub voce.*

3. ASV, Senato 111, Secreta, Palma 7 (1605–7), 27 March 1606 (received 4 April 1606).

4. ASV, Senato 111, Secreta, Palma 7 (1605–7), 19 April 1606. For similar cases, see De Vivo, *Patrizi, informatori, barbieri*, 74–78.

5. ASV, Senato 111, Secreta, Palma 7 (1605–7), 30 April and 11 May 1606.

6. See Armando Petrucci, *Public Lettering: Script, Power and Culture* (Chicago: University

of Chicago Press, 1993) for a wide-ranging analysis of the changing role of public writing, from ancient epigraphy to postwar graffiti. See also Andrew Pettigree, *The Book in the Renaissance* (New Haven: Yale University Press, 2010), chapter 16.

7. ASV, Senato III, Secreta, Palma 7 (1605–7), 19 April 1606, with a copy of a letter from the Capuchin provincial to the vicario in Palma confirming their obedience, sent 25 April 1606.

8. ASV, Senato III, Secreta, Palma 7 (1605–7), 14 May 1606.

9. ASV, Senato III, Secreta, Palma 7 (1605–7), 15 May 1606, Gussoni to Donà, with a letter enclosed from Fra Amadeo Veronese to the vicario of the Capuchins in Palma, dated 9 May 1606.

10. ASV, Senato III, Secreta, Palma 7 (1605–7), 16 May 1606.

11. Donà had been part of the original reconnaissance mission of 1593, keeping a detailed travel journal. See the excellent account in Deborah Howard, *Venice Disputed: Marc' Antonio Barbaro and Venetian Architecture, 1550–1600* (New Haven: Yale University Press, 2011), 199–200.

12. ASV, Senato III, Secreta, Palma 7 (1605–7), 11 May 1606.

13. ASV, Senato III, Secreta, Palma 7 (1605–7), 22 April 1606.

14. Howard, *Venice Disputed*, chapter 6.

15. Ibid., 207–9.

16. See, for example, his letter to Galileo of 18 October 1602, where he passes on the news of the taking of Buda by the Turks; *OG* 10:96–97. It is worth noting here that the Palazzo Sagredo in Castello—where Sagredo lived until he moved with his father into the Procuratie Vecchie at San Marco in 1612, and which was left vacant at the death of doge Leonardo Donà—was next door to the official residence of the Roman nunzio, the former Palazzo Gritti, donated by the Venetians to Sixtus V in 1586 in exchange for the Palazzo Venezia in Rome. This was a major information-gathering center.

17. Correr, Cod. Cic. 3014, n.37, f.1r–2v. On Tiepolo, elected to the position of *primocerio* in 1603, see Emmanuele Antonio Cicogna, *Delle inscrizioni veneziane* (Venice: Giuseppe Picotti, 1830), 3:91, and Gaetano Cozzi, "Note su Giovanni Tiepolo, primicerio di San Marco e patriarca di venezia: L'unità ideale della chiesa veneta," in *Chiesa, società e stato a Venezia: Miscellanea di studi in onore di Silvio Tramontin nel suo 75 anno di età,* ed. Bruno Bertoli (Venice: Studium cattolico Veneziano, 1994), 121–50; and, on the office of the *primocerio,* Gaetano Cozzi, "Gius-patronato del doge e prerogative del primicerio sulla Cappella Ducale di San Marco (secoli XVI–XVIII): Controversie con i procuratori di San Marco de supra e i patriarchi di Venezia," in *Atti dell'Istituto veneto di scienze lettere ed arti* 151 (1992–93): 1–69. Sagredo seems to have been well-connected at this time; in December 1606 he sent game from Palmanova to the new doge, Leonardo Donà, to celebrate his election. Donà preserved Sagredo's letter, along with those of a few other well-wishers. It is now in Archivio Donà delle Rose, Fondamento nuove, Venice, n.XXII, b) Lettere di Gioan Francesco Sagredo (Palma), Giovanni Renier, Gioan Francesco Dol-fin, etc riguardanti l'invio a Leonardo e a Nicolò Donà di selvaggina per il banchetto elettorale del Doge. Dicembre 1606 (unconsulted; reference supplied by Prof. R. Zago).

18. Correr, Cod. Cic. 3014, n.37, f.1r–2v.

19. Wotton to Salisbury, 22 April 1606. Smith, *Life and Letters,* 1:345.

20. Wotton to Salisbury, 25 August 1606. Smith, *Life and Letters,* 1:359.

21. Wotton to Salisbury, May 1606. Smith, *Life and Letters,* 1:351.

22. Wotton to Salisbury, 31 July 1609. Smith, *Life and Letters,* 1:463.

23. *CSP, Ven.,* 11:494, doc. 917.

24. "Ottavio Baldi" (Wotton's alias) to James I, 3 November 1606. Smith, *Life and Letters*, 1:366–67.

25. *CSP, Ven* 11:9.

26. Wotton to Salisbury, 25 August 1606. Smith, *Life and Letters*, 1:359.

27. On the episode, see Smith, *Life and Letters*, 1:480, and Bouwsma, *Venice and the Defense of Republican Liberty*, 505.

28. Smith, *Life and Letters*, 1:497.

29. Wotton to Salisbury, 13 March 1610. Smith, *Life and Letters*, 1:485.

30. A similarly detrimental decontextualization has been suffered by Paolo Sarpi's famous letter announcing the arrival of the first spyglasses in Venice. After a warning concerning the dangers of interception and misinterpretation, Sarpi writes: "In Italy there's nothing new: only that spyglass for seeing faraway things, which I admire greatly for the beauty of its invention and the skill of its manufacture, but which I consider worthless for use in war either at sea or on land." So far, so good, but the letter then continues:

> The Turkish armada has finally left to great pomp. It contains sixty galleys, with one large one, and as many other boats. It is heading for Alexandria in Egypt, I think, to make sure of the *cassnà*, that is, the tax they take from there to Constantinople. On the way back they'll stop at Morea, and from there go to Sicily or Calabria, which may not happen, but in any case it will be rather light. Bohemian affairs are reported here in a very bad state, and even so in Rome they don't think about it, both because they're faraway things and because they take away obsequies and adoration, but not money, which is the only thing left of value. The King of England's book has arrived [. . .] (Sarpi to Francesco Castrino, 21 July 1609, *Scelte lettere inedite di frà Paolo Sarpi* [Capolago: Elvetica, 1833], 72–74).

In Sarpi's letter, the telescope is both a piece of information and a (poor) instrument for making information: Sarpi's own letter system immediately provides precisely the kind of news he doubts the telescope capable of producing. The best accounts of the telescope as an essentially political device are Reeves, *Galileo's Glassworks*, and Massimo Bucciantini, Michele Camerota, and Franco Giudice, *Il Telescopio di Galileo: Una storia europea* (Turin: Einaudi, 2012). On Sarpi's letter, see Mario Biagioli, "Did Galileo copy the telescope? A 'new' letter by Paolo Sarpi," in *The Origins of the Telescope*, ed. Albert Van Helden, Sven Dupré, Rob van Gent, and Huib Zuidervaart (Amsterdam: KNAW, 2010), 203–30.

31. See *DBI* entry, *sub voce*, and Pietro Savio, "Per l'epistolario di Paolo Sarpi," *Aevum* 16 (1942): 17–21, for an excellent collection of primary documents. Menini also had an international audience; his poetry was sent to Isaac Casaubon by Domenico Molino, along with Galileo's recently published *Sidereus nuncius*, in April 1610, just before Casaubon moved from Paris to London. See *Catalogue of Manuscripts in the British Museum, New Series*, vol. 1, part 2, *The Burney Manuscripts* (London: 1840), 122, describing British Library, Burney mss. 367. Erasmo di Valvasone included some of Menini's poems and a reference to Menini's lost discourse defending the poetic treatment of angels in the dedicatory letter of his *Angeleida* (1590).

32. On Groslot de l'Isle's role in information networks, see Filippo de Vivo, "Paolo Sarpi and the Uses of Information in Seventeenth-Century Venice," in *News Networks in Seventeenth-Century Britain and Europe*, ed. Joad Raymond (London: Routledge, 2005), 35–50.

33. Ottavio Menini, *Carmina: Ad res potissimum Gallicas, Venetas, & Romanas pertinentia, varijs temporibus scripta. Eiusdem Panegyricus serenissimo principi Donato, & excelso Senatui*

Veneto dictus. Eiusdem De Pace, oratio. Eiusdem De Nece Henrici 4. & de Inauguratione Ludouici 13. Ad Proceres Gallos, oratio (Venice: 1613), 138–40.

34. The sheet, as Favaro remarks, bears Galileo's address in Padua and was sealed with Sagredo's signet ring. This may mean that Sagredo and perhaps other guests at the Barrozzi banquet were given the text of the poem before its oral delivery on 1 August 1608. It is located at BNCF, Ms. Gal. 316, ff. 340r–341v. No letters exist from Sagredo to Galileo between 30 April 1608 and 28 October 1609, when he first wrote from Aleppo.

35. Sarpi, *Lettere ai Protestanti*, 1:25, n.VIII (al Groslot de l'Isle, 5 August 1608) and 2:130 (interview with Christoph von Dohna, 23 August 1608).

36. Favaro, Antonio, "Ancora a proposito di Giovanfrancesco Sagredo," 510. Favaro quoted at length from the Sarpi letter in his 1902 article "Giovanfrancesco Sagredo e la vita scientifica in Venezia al principio del XVII secolo," *Nuovo archivio veneto* 4 (1902): 328, and noted that it had been published in 1673. On Favaro's troubled relationship with Sarpi's correspondence, see Biagioli, "Did Galileo copy the telescope?"

37. "*Ad Hilaritatem* / Huc alacrem fer, Diva, pedem, quae pectore curas / Et fronte nubes discutis. / Diva, veni laetamque domum festosque Penates / Barocii lubens adi. / Te lectus Procerum coetus, flos ordinis alti, / Te sceptra, fasces, purpura / Paene sua oblita et solita gravitate carentes / Ad plena poscunt pocula. / Sextiles veniam dant istam nempe Calendae, / Qua nulla lux felicior, / Nulla supercilii minus alti, nulla severis / Minus revincta legibus. / Luce bona bona vina decet potare: Falerni / Minister huc ades, puer: / Porge, puer, pateram spumantem nectare dulci / Bis, ter, quaterque et amplius; / Ah, procul hinc absint lymphae, ac tetra venena / Mensis ab istis exulent. / Porge, puer, citharam: cithara tenuisse iuvabit / Aures virorum principum. / Heroes salvete boni, mea Numina: Vobis / Serenus hic eat dies. / Hic redeat vobis, centum redeuntibus annis, / Et usque et usque laetior. / Heroes soliti magnas dare iura per urbes, / Prisci Catonis aemuli, / Nunc genio indulgete: brevis sic postulat hora: / Ad magna cras redibitis. / Exhilara ante alios te, coetus huius ocelle / Sagrede: tu quidem brevi / Littore ab hoc solves, longum trans aequor iturus, / Quo cara mittit Patria, / Et varios inter populos ac Barbara regna / Custos futurus civium. / At tibi felices cursus laetosque recursus / Praestent amica sydera, / Et post solerti tractata negocia dextra, / Virtute post partum decus, / Te te restituant opibusque et honoribus auctum / Cupidis tuorum amplexibus. / Nos hic interea memori te mente fruemur, / Imago nec tua effluet / Pectoribus nostris; unquam non effluet iste / Decor leporque amabilis. / Seria nec tantum tua nos meminisse iuvabit, / Sed et iocosa et ludicra. / Dum tamen hinc aberis, quis posthac ludet hiantes / Illos rapaces alites, / Quos procul hinc altos male credula turba Quiritum / Tot candidos cycnos putat? / Nomine quis ficto matronae divitis, illas / Dulces tabellas exprimet? / Et captatores capiet, pari et arte docebit / Artes dolosas vincere? / O redeas propere, redeas felicibus Austris! / Nam, te recepto sospite, / Haec eadem tecum repetent carchesia rursus / Laeti sodales et tuas / Attollent caelo laudes atque inclita facta, / Quin et puella candida / Te reducem excipiet plaudentibus obvia palmis / Praesensque ceu numen colet: / Illa puella, suas artes cui cessit Apollo, / Quae spirat omnes gratias, / Illa puella, dolis sancta sub imagine tectis / Decepta quae quondam fuit, / Quaeque modo illorum insidias, fallacibus hamis / Qui credulam ad se attraxerant, / Conscia iam veri retegit profertque sub auras / Mire venusto carmine. / Sed quo abeo? Da vina, puer: iuvat usque madentem / Diem bibendo ducere. / Ah, ne Sol propera: sit lux haec longior anno, / Nec ipse surgat Hesperus." My profound thanks to Kathleen Coleman for help with the translation.

38. On the term "chietine," meaning "devotee," see Pavone, *Le Astuzie dei Gesuiti*, 158n37.

39. *OG* 10:203.

40. *OG* 10:204.

41. *OG* 12:454.

42. Sarpi to Antonio Foscarini, 27 May 1608. Pietro Savio, "Per l'epistolario di Paolo Sarpi (cont.)," *Aevum* 11 (1937): 31.

43. Ibid., 31–32.

44. Marciana, Cod. It., X, 220 (6409), contains a mid-nineteenth century catalogue of Girolamo Contarini's manuscripts, where Sagredo's exchange, as well as the manuscript now preceding it in the Marciana—a dialogue between a priest and the Gobbo (now Marciana, Cod. It., X, 189 (7217) "L'Ombra di Aristofane Atheniese")—is described. It is interesting to note the context of these works in the Contarini catalog, as several of them may have come from Gianfrancesco Sagredo: 121r ("Raccolta di varie Scritture concernenti li Padri Gesuiti"), f.123r ("Raccolta di varie composizioni: Cronica Veneta della controversia contro Paulo V), f.123v ("L'Ombra di Aristofane Ateniese Discorsi di F. Manzino, e del Gobbo di Rialto"), f.124r ("Lettere di M. Rocco Berlinzone"), and f.125v ("Istoria dell'Interdetto fulminato da Paolo V contro la Republica").

45. On Contarini, see both Gaetano Cozzi, *Il doge Niccolò Contarini: Ricerche sul patriziato veneziano agli inizi dei Seicento* (Venice and Rome: Istituto per la collaborazione culturale, 1958) and Cozzi's entry in *DBI, sub voce.*

46. See Favaro, "Giovanfrancesco Sagredo," Doc. VII.

47. Barisone entered the Society on 25 March 1576, and became a professor at Padua in 1596. For details, see the record in ARSI, Ven. 38, Cat. Trien. 1603–19, f.95r. He became rector of the College at Ferrara on 25 February 1606 (ARSI, Ven. 5, 1600–1606, f.446v). He died on 7 September 1623 (ARSI, Hist. Soc. 42, Defunti 1557–1623, f.14r). As rector in Ferrara, he was one of the recipients of the bull of excommunication, sent out by the general with instructions to publish it immediately. See ARSI, Ep. Gen. Ven 5II, 457r (19 April 1606). Other copies were sent to the rectors of Parma, Forlì, Mantova, Piacenza, Modena, and Bologna, as well as to the colleges of Venice, Brescia, Padova, Verona, and Vicenza. A description of the process of the Interdict from the point of view of the Ferrarese Jesuits and an account of the arrival of refugee Jesuits from the Veneto to Ferrara is at ARSI, Ven 127, ff.47v–48r. In 1606 the Ferrara College had twenty-one Jesuit priests and twenty coadjutors, each with an income of one thousand scudi a year (ARSI, Ven. 38, Cat. Trien, 1603–19, f.125r). Barisone was reprimanded by the general in January 1607 for leaving the corpse of the Marquis of Massa below the altar of the sacristy in the Jesuit church at Ferrara, apparently as a favor in exchange for his family endowing a new chapel. (ARSI, Epistolae Gen. Veneta 6, 1607–16, f.6v (31 January 1607); f.7r (no date, between 31 January and 3 February 1607; f.13r, 24 March 1607).

48. Marciana, Cod.It., x.188 (7216), f.12r (12 April 1608)

49. It is worth noting that Gianfrancesco had a sister named Angela who married a Contarini, and a sister-in-law named Cecilia. The 1612 *Vocabolario* of the Crusca cites "Berlinzone" as one of several examples from the *Decamerone*, day 8, story 9, as an "empty, made-up name, as a prank" (*nome vano, e finto, per baia*), or as a "name made for a joke, to mean something incomprehensible" (*nome formato per ischerzo, e per dir cosa, che non s'intenda*). See entries for "Ciancianfera," "Narsia," "Scalpedra," and "Semistante."

50. A draft of Contarini's will is included at Marciana, Cod.It., x.188 (7216), f.23r–24v.

51. Giuseppe Cappelletti, *I Gesuiti e la Republica di Venetia* (Venice: Grimaldo, 1873), 200–201.

52. Ibid., 201.

53. ARSI, Ven. 6, f.68v (5 July 1608).

54. Marciana, Cod.It., x.188 (7216), f.66r, dated 12 July 1608.

55. 'Sarpian' authorship is extended (and perhaps overextended) to many other political

texts in Luigi Lazzerini, "Officina sarpiana: Scritture del Sarpi in materia di Gesuiti," *Rivista di storia della Chiesa in Italia* 58 (2004): 29–80. Despite the methodological looseness in some of these attributions, we would do well to start thinking about collective authorship for some early modern texts.

56. Sarpi, *Lettere ai Protestanti*, 2:130 (interview with von Dohna, 23 August 1608, misidentified in the index as "probably Paolo Barisoni, Paduan"), cited but unidentified in Lazzerini, "Officina sarpiana," 29.

57. "Scritture et avisi havuti da diverse persone concernenti le insidiose machinationi et male actioni de Padri Gesuiti verso questa Serenissima Republica." The entire volume is printed in Cappelletti, *I Gesuiti e la Repubblica*. For a description of ASV, consultori in iure 454, see Corrado Pin, "Tra religione e politica: Un codice di memorie di Paolo Sarpi," in *Studi politici in onore di Luigi Firpo*, vol. 2, eds. S. Rota Ghibaudi and F. Barcia (Milan: FrancoAngeli, 1990). The copy of the letter is described at 155–56n20.

58. Vittorio Frajesi, *Sarpi Scettico: Stato e Chiesa a Venezia tra cinque e seicento* (Bologna: Il Mulino, 1994), 213.

59. Cicognara described the volume in his entry on S. Giorgio Maggiore, where he provided a partial bibliography of the Interdict. See *Delle Iscrizione Veneziane* 4:440n10. Sir Henry Wotton explained the motives and mechanism behind the dossier:

> They have here newly concluded in Senate the process of the Jesuits, who, with great unity of consent, are banished out of all their towns and territories for ever; and because their skill is known in working into states out of which they have been ejected, there are here decreed three provisions against the possibility of their return: first, that they must pass *le strettezze delle balle* (as we term it) in the College, and have the whole twenty-seven balls in favour, before the proposition by admitted into the Senate. Then, that in Senate they must carry five-sixth of the voices, which are 150 of 180. And thirdly, that in both places (before proceeding to the ballotation) the whole process must be read, both of that matter which hath been hitherto alleged against them, and shall be hereafter. And this last provision was proposed by the Duke himself, as well for irritation as for delay [Wotton to Salisbury, 23 June 1606. Smith, *Life and Letters*, 1:354].

60. Ottavio Baldi (Henry Wotton) to James I, 20 March 1609. Smith, *Life and Letters*, 1:449.

61. Marciana, Cod.It., x.188 (7216), f.68v.

62. Sebastian died on a crusade to Morocco in 1578. See Dauril Alden, *The Making of an Enterprise: The Society of Jesus in Portugal, Its Empire, and Beyond, 1540–1750* (Stanford, CA: Stanford University Press, 1996), chapter 4.

63. Marciana, Cod.It., x.188 (7216), f.84v.

64. On Querini, see below.

65. Cappelletti, *I Gesuiti e la Repubblica*, 141 (doc. 82).

66. This is an emerging field; look out for future work by Paula Findlen and Alex Marr on humor and ingenuity.

Chapter Five

1. Sagredo certainly wrote other letters from Syria, but all those that survive are on official business. In addition to those discussed below, see, for example, *CSP, Ven.* 9:847: Simone Con-

tarini, Venetian ambassador in Constantinople, to the doge and Senate, referring to "a letter from the Consul at Aleppo," 3 April 1610.

2. ASV, Consultori in Iure 453.

3. Presumably Sarpi's friend Fulgentio Micanzio, not Fulgentio Manfredi.

4. *OG* 10:242–43.

5. Schaffer, "Newton on the Beach"

6. On the figure of Peiresc, see, above all, Peter Miller, *Peiresc's Europe: Learning and virtue in the seventeenth century* (New Haven: Yale University Press, 2000).

7. The bibliography on the globalization of Jesuit science is now vast. See, by way of introduction, Steven J. Harris, "Confession-Building, Long-Distance Networks, and the Organization of Jesuit Science," *Early Science and Medicine* 1 (1996): 287–318; Mordechai Feingold, ed., *Jesuit Science and the Republic of Letters*, (Cambridge, MA: MIT Press, 2003); Mordechai Feingold, ed., *The New Science and Jesuit Science: Seventeenth Century Perspectives* (Boston, MA: Kluwer Academic Publishers, 2003); Luís Miguel Carolino and Carlos Ziller Camenietzki, eds., *Jesuítas, ensino e ciência, séc. XVI–XVIII*, (Casal de Cambra: Caleidoscópio, 2005); Ugo Baldini, "The Jesuit college in Macao as a meeting point of the European, Chinese and Japanese mathematical traditions: some remarks on the present state of research, mainly concerning sources, 16th–17th centuries" in *The Jesuits, the Padroado and East Asian science, 1552–1773: History of mathematical sciences: Portugal and East Asia III*, ed. Luís Saraiva and Catherine Jami, (Singapore: World scientific, 2008); Florence C. Hsia, *Sojourners in a Strange Land: Jesuits and their Scientific Missions in Late Imperial China* (Chicago: University of Chicago Press, 2009); Andrès I. Prieto, *Missionary Scientists: Jesuit Science in Spanish South America, 1570–1810* (Nashville: Vanderbilt University Press, 2011) and my own "Science and the Counter-Reformation," in *The Ashgate Companion to the Counter-Reformation*, ed. Mary Laven, Alexandra Bamji and Geert Janssen, (Aldershot: Ashgate, 2013).

8. For an interesting corrective to this view, which also helps open up the curiously neglected field of Mediterranean science, see Ben-Zaken, *Cross-cultural Exchanges*.

9. *OG* 10:96.

10. *OG* 10:262.

11. *OG* 12:335.

12. *OG* 11:172.

13. Sarpi, *Lettere ai Protestanti*, 1:184. Sagredo delivered his first oral *relazioni* on Syria the morning of 4 July 1611 (ASV, Collegio, Relazione di Ambasciatori, b.31, Soria, 7a); on that occasion he took the opportunity to "bore" the Senate with a reading of the correct version of a paper of economic policy he had sent on 17 November 1609, which he claimed had circulated in Venice in bad copies during his absence. A second *relazione* (ASV, Collegio, Relazione di Ambasciatori, b.31, Soria, 6a) was presented on 11 January 1612 and delivered 15 May 1612, which in turn promised a third *relazione* on Persia, now lost. The two *relazioni*, both fair copies with autograph corrections, were printed in Guglielmo Berchet, *Relazioni dei Consoli Veneti nella Siria* (Turin, Paravia, 1866), 130–37 and 138–56, and also by Favaro in "Giovanfrancesco Sagredo." There is some evidence of scribal publication of the *relazioni*, or at least of Sagredo's family retaining a copy: Correr, Mss. P.D. C.2193, III, a 1763 inventory of books ("Inventario e stima de' Libri ritrovati in essere di rag[io]ne dell'Eredità del fù N.H.M. Girardo Sagredo Proc.r di S. Marco"), includes at n19 "Relazione d'Aleppo et altre Scritture. in fol. patito dall'Acqua, e tarmato" and, more definitively at n198, "Dispacci Publici de'Residenti in Alessandretta, d'Aleppo, ed altri luoghi del 1608, 1609, 1610 1611 in fol." The Leopold Von Ranke Manuscript Collection at

Syracuse contains a copy of Sagredo's second *relazione* at Ms. 402, G, "Relatione del nob. homo s. Zuan Francesco Sagredo, ritornato console dalla Soria, presentata nell'ecc. Collegio. 1611. 16 gennaro," bound in a chronologically arranged volume of Venetian *relazioni* from Syria. See Edward Muir, *Leopold Von Ranke Manuscript Collection of Syracuse University: The Complete Catalogue* (Syracuse, NY: Syracuse University Press, 1983), 250.

14. For a detailed reconstruction of the part of Sarpi's archive transferred upon his death to the State archives, see Corrado Pin, "Le Scritture pubbliche trovate alla morte di fra Paolo Sarpi nel convento dei Servi," *Memorie della Accademia delle Scienze di Torino* 2 (1978): 311–69.

15. The cost of setting up an information order was high. Wotton's operations cost his patrons at least two hundred pounds every three months. Smith, *Life and Letters*, 1:351, 359.

16. "Io non hò osservati li pianeti medicei; ben essendo in Soria osservai le stelle medicee col primo Instrum[en]to che io hebbi, anzi avanti, che io l'havessi restava [*sic*] in grande aspetat[ion]e per osservare le istesse costellationi, ch'à punto ella hà osservato: onde leggendo poi il Sidereus Nuncius restai con qualche maraviglia d'havere incontrato così puntualmente la istessa parte del Cielo. Se mi sara da lei mandato le sue osservationi de sudetti pianeti sara cagione, che io li osservaro." *OG* 11:314 (original punctuation restored from BNCF, Mss Gal. 89, f. 16v).

17. Paolo Preto, *I servizi segreti di Venezia: Spionaggio e controspianaggio ai tempi della Serenissima* (Milan: Il Saggiatore, 1994), 228n102. For intercepted letters in particular, citing the example of Sagredo on 293, see 293–300.

18. ASV, Inquisitori di Stato, 516, ff.30r–53v, with Sagredo's signet seal. Note of registration: "Manda la copia di alcune 1[ette]re di Persia da lui intercette. Et anco di altro fogli. ~~Dice di mandar altre 1[ette]re in Idioma, et carattere Persiano, ma queste no[n] si sono trovato nel piego~~ <in carattere, et Idioma Persiano> ~~fù risposto 21 Giugno 1610.~~ Fù risposto p[er] C[onsiglio] di X. 21 Giugno 1610." Further copies, with other material, are located at ff.54r–76v.

19. For the background to this scheme, see Giorgio Rota, "Safavid Envoys in Venice," in *Diplomatisches Zeremoniell in Europa und im Mittleren Osten in der Frühen Neuzeit*, ed. Ralph Kauz, Giorgio Rota, and Jan Paul Niederkorn (Vienna: Verlag der Österreichischen Akademie der Wissenschaften, 2009), 213–45; Giorgio Rota, *Under Two Lions: On the Knowledge of Persia in the Republic of Venice (ca. 1450–1797)* (Vienna: Verlag der Österreichischen Akademie der Wissenschaften, 2009); and Giorgio Rota, "Safavid Persia and Its Diplomatic Relations with Venice," in *Iran and the World in the Safavid Age*, ed. Willem Floor and Edmund Herzig (London: I. B. Tauris, 2012), 149–60.

20. ASV, Inquisitori di Stato, 516, f.33v.

21. Ibid. ff.37r–52v. On the Shirley brothers, see, most recently, the entries in ODNB and Sanjay Subrahmanyam, *Three Ways to be Alien: Travails and Encounters in the Early Modern World* (Waltham, MA: Brandeis University Press, 2011).

22. ASV, Inquisitori di Stato, 516, ff.73r–74v (deciphered, translated); other copies at ff.65r–66v (deciphered, translated), ff.17r–18v (ciphered), ff.25r–26v (ciphered), and 27r–29v (original copy, ciphered, with envelope at 28rv).

23. This information relies on Rudi Matthee's excellent account of de Gouvea, *sub voce*, in the *Encyclopedia Iranica*, ed. Ehsan Yarshater, (Costa Mesa: Mazda, 2003) 11:177–79. For the broader story, see Joan-Pau Rubiés, "A Dysfunctional Empire? The European Context to Don Garcìa de Silva y Figueroa's embassy to Shah Abbas," in *Estudos sobre Don García de Silva y Figueroa e os "Comentarios" da embaixada à Persia (1614–1624)*, ed. Rui Loureiro and Vasco Resende (Lisbon: Centro de História de Além-Mar, 2011), 85–133, which builds on the classic study

by Niels Steensgaard, *Carracks, Caravans and Companies: The Structural Crisis in the European-Asian Trade in the Early 17th Century* (Lundun: Studentlitteratur, 1973).

24. ASV, Inquisitori di Stato, 516, ff. 3r–11r. The original ciphered version is at 27r–29v, with copies at 13r–16v and 21r–24v.

25. ASV, Esp. Prin. 18, 2 September 1609 (unpaginated). The letter was presented by Ṣafar on 22 January 1610. Printed in Berchet, *La repubblica di Venezia e la Persia*, 201–3.

26. ASV, Inquisitori di Stato, 516, ff.67r–69v.

27. Ibid. f.71r; minutes of the letter are at ASV, Consiglio di Dieci, Parte secrete, reg. 15, f.94r, along with a note recording the permanent transfer of the entire set of documents from the Council of Ten's archives to those of the Inquisitors of State, under seal.

28. ASV, Consiglio dei X, Parte secrete, reg. 15, f.94 (21 June 1610).

29. ASV, Inquisitori di Stato, 180. Soria, f.1r.

30. ASV, Inquisitori di Stato, 516, ff.78v. On the theory and practice of Venetian codes, see Preto, *I servizi segreti di Venezia*, 268–279.

31. ASV, Diritti e Canoni, Capo XXIX, Cap. 2584, Art.3, ff.1r–31v.

32. Ibid., f.30r.

33. The literature on early modern conversion is vast. See, most suggestively, Natalie Zemon Davis, *Trickster Travels: A Sixteenth-Century Muslim between Worlds* (New York: Hill & Wang, 2006).

34. ASV, Diritti c Canoni, Capo XXIX, Cap. 2584, Art.3, f.8r.

35. ASV, Quarantia Criminal, Processi, b.127, n.183, containing two near identical copies, both signed and sealed by Sagredo and sent to Venice on 26 February 1610 and 10 April 1610. These dossiers may be supplemented by the French supplications to the doge recorded in ASV, Esp. Princ. 18, 14 October 1609, containing copies of documents written between January and June 1609.

Chapter Six

1. See, for example, Edward Rosen, "The Title of Galileo's *Sidereus Nuncius*," *Isis* 41 (1950); Stillman Drake, "The Starry Messenger," *Isis* 49 (1958): 346–47; and, for an attempted closure of the debate, F. Russo, "Note sur la traduction du titre de l'ouvrage de Galilée, *Sidereus nuncius*," *Revue d'histoire des sciences* 20 (1967): 67–69.

2. Within the main text, the drop title, printed in late January, is "Astronomicus Nuncius observationes recens habitas novi perspicilli beneficio . . . continens atque declarans" (Astronomical Message that contains and explains recent observations made with the aid of a new spyglass), while the running title is peculiarly related to this: "Observat[iones] Sidereae / Recens Habitae" (Sidereal observations recently carried out).

3. The words mean, more or less, "herald," "spy," and "message" or "messenger."

4. Favaro lists the 1590 Calepino, *Dictionarium in quinque linguis,* published by Domenico Farri, as being in Galileo's library: a folio work in three volumes with additions by Paolo Manuzio, which were first introduced in 1558. We need not take the specificity of Favaro's claim too seriously, as most of his identifications of editions were highly educated guesses. In the following section, I used the 1590 Griffo edition. Antonio Favaro, "La libreria di Galileo Galilei descritta ed illustrata," *Bullettino di Bibliografia e di Storia delle Scienze Matematiche e Fisiche* 19 (1886): 277n315.

5. Alexander Scot, *Universa Grammatica Graeca* (Lyon: 1594).

6. "Κηρυξω, per praeconem denuntiasse," Scot, *Universa Grammatica Graeca*, 522. On the use of trumpets at early modern state rituals, see Stephen Rose, "Trumpeters and Diplomacy on the Eve of the Thirty Years' War: The 'Album Amicorum' of Jonas Kröschel," *Early Music* 40 (2012).

7. "Nunciatio, onis, [Annonciatione. Gal. Nouuciation, annoncement Hi. Obra de traher mensaie.] Αγγελια, denunciandi actus. Calepino, *Dictionarium, sub voce.*

8. "Index, cis, [Demonstratione, indicio. Gal. Monstreur ou monstre. His. Descubridor del secreto,]" Calepino, *Dictionarium, sub voce.*

9. "Nuncius, cij, vel potius nuntius, [Messo, messagiere, o ambasciatore Ga. Messager, ou message Hi Mensaiero ò mensare.], αγγελος, αγγελια, qui nuntiat, & quod nuntiatur." Calepino, *Dictionarium, sub voce.*

10. *Mercurii gallobelgici: Siue Rerum in Gallia & Belgio potissimum: Hispania quoque, Italia, Anglia, Germania, Polonia, vicinisque locis ab anno 1588, vsque ad Martium anni præsentis 1594, gestarum, nuncii tomus primus* (Cologne: 1598).

11. Probably written before 1597. See John Donne, *The Epigrams, Epithalamions, Epitaphs, Inscriptions, and Miscellaneous Poems*, ed. Ted-Larry Pebworth, Gary A. Stringer, Ernest W. Sullivan II and William A. McClung. Volume 8 of *The Variorum Edition of the Poetry of John Donne*, ed. Gary Stringer, (Bloomington: Indiana University Press, 1995).

12. See, for example, *Bando che nessuno possa scriver Lettere d'Avisi, Senza Licenza di Monsignor Governatore* (Rome: Nella Stamperia della Camera Apostolica, 19 September 1602). Copy consulted: Casanatense, Per. Est, 18.3 (1600–1605), ff. 275 and 278).

13. "Avviso" is a loan word from Portuguese. Even in English the word "advice" itself could mean, in the seventeenth century, something like "information." See, for example, Henry Wotton to Sir Arthur Throckmorton, 8 May 1611, "We have advice out of Germany," Smith, *Life and Letters*, 1:507.

14. See, for example, Agesilao Marescotti's *Aviso sicuro contro il mal fondato auiso del signor Antonio Quirino senator Veneto*, from Bologna; and, from the same city, Antonio Possevino's pseudonymous *Risposta di Teodoro Eugenio di Famagosta, all'Auiso mandato fuori dal signore Antonio Quirino senatore Veneto* (1606), reprinted in Bologna and Viterbo the next year; and the same author's *Risposta del sig. Paolo Anafesto all'auuiso del sig. Antonio Quirino*, published in Camerino and Bologna in 1607 (note the proliferation of pseudonyms). For a general description of the positions assumed in these works, see Bouwsma, *Venice and the Defence of Republican Liberty*, chapter 8.

15. *OG* 10:76; 19:222, 224, and 226; and 2:560. A previously unnoticed autograph note on the flyleaf of Galileo's heavily annotated copy of Capra's plagiarized *Usus* says, "Excuse Signor Querini with Dr. Contarini if I have not written back to him," [Scusare il S. Quirini co[n] Dott. Contarini se non li ho rescritto.] (BNCF, Ms. Gal. 40, 3v)—presumably meaning that Galileo's own abusively annotated copy circulated as evidence between the rectors in Capra's trial, and that nonwritten persuasion was also admissible.

16. Kepler called the work a "libellus," implying not only that it was a pamphlet, but perhaps also that it was defamatory or incendiary (*OG* 3:320, 10:105).

17. See, for example, [Antonio Possevino?], *Relatione della Segnalata et come miracolosa Conquista del Paterno Imperio.* [. . .] *Raccolta da sincerissimi avvisi per Barezzo Barezzi* (Venice: 1605) (Florence: 1606).

18. On the scribal publication of *relazioni*, see de Vivo, *Information and Communication in Venice*, 57–70.

19. Filippo De Vivo, "How to Read Venetian Relazioni," *Renaissance and Reformation*, 34 (2011); and, in the same volume, Andreas Motsch, "Relations of Travel, Itinerary of a Practice."

20. See Johannes Kepler, *Dissertatio cum Nuncio sidereo / Discussion avec le Messager céleste*, ed. and trans. Isabelle Pantin (Paris: Les Belles Lettres, 1993), xxxvi–xlix, for the relationship between Horky and Capra, as well as Bucciantini, *Galileo e Keplero*, 181.

21. 9 August 1610. On Welser, see Giuseppe Gabrieli, "Marco Welser Linceo Augustano," *Rendiconti della R. Accademia Nazionale dei Lincei: Classe di scienze morali, storiche e filologiche* 13 (1937); and Pietro Redondi, "I fondamenti metafisici della fisica di Galileo," *Nuncius: Annali di storia della scienza* 12 (1997).

22. Despite my discovery that the supposed autograph proof copy of the *Sidereus nuncius* owned by Martyan Lan, known as "SNML," is a modern fake, much (but not all) of the evidence assembled by Paul Needham in *Galileo Makes a Book*, vol. 2 of *Galileo's O*, ed. Horst Bredekamp (Berlin: Akademie Verlag, 2011), concerning the composition and print history of the *Sidereus nuncius* still holds true. "SNML" was first authenticated by Horst Bredekamp in *Galilei der Künstler: Der Mond. Die Sonne. Die Hand* (Berlin: Akademie Verlag, 2007). Much of that book and volume 1 of *Galileo's O*, *Galileo's Sidereus Nuncius*, ed. Irene Brückle and Oliver Hahn, must now be heavily revised or discarded.

23. ASV, *Riformatori allo studio*, 168. On Tommaso Baglioni, see A. Cioni's entry in the *DBI*, which, however, concentrates mainly on the later career and eighteenth-century *fortuna* of the Baglioni house.

24. Bredekamp provides a fine discussion of the image in *Galilei der Künstler*, 116–19.

25. The next time the *P* appears is three years later, in Roberto Meietti's 1603 edition of Francisco Toledo's *Commentarij in Epistolam beati Pauli apostoli ad Romanos*, where it is used twice. In 1606 it again appears twice in Meietti's edition of Giacomo Sbrozzi's *De vicarii episcopi officio, et potestate tractatus*: in the dedicatory letter at +2r and at the start of the appendices after page 370 at Aa2r. This book was also issued with a variant title page ascribing it to Giovanni Battista and Giovanni Bernardo Sessa, though the printer's colophon belongs to Meietti. Then, in 1607, it appears in two more editions: in Benito Pereira's *Centum octoginta tres disputationes selectissimae super libro Apocalypsis Ioannis Apostoli*, published by Antonio Leonardi in Venice, on page 180; and again in Meietti's Venetian edition of Giovanni Marsilio's *Seconda Parte dell'Essame*, on page 2.

26. Antonio Favaro "Intorno alla licenza di stampa del Sidereus nuncius di Galileo Galilei," *Rivista delle biblioteche* 18–19 (giugno-luglio 1889): 98–103. The only attention previously paid to this document focused on Favaro's uncharacteristic mistakes. These he partially and silently emended in volume 19 of Galileo's *Opere*: first, he misdated the document, reading the open loop of the final digit of "1609" as a "7." He also misread "Bellarino" as a mistranscription of "Bellarmino," even though Bellarmine never printed any book with such a title. This led him to posit a convoluted claim that Galileo had intended in 1607 to publish a book entitled *Astronomica denuntiato ad astrologos*—a book he never got around to writing, but whose license he recycled three years later for the *Sidereus nuncius*. See Antonio Poppi, "Una implicita ritrattazione di Antonio Favaro," in his *Ricerche sulla teologia e la scienza nella Scuola padovana del Cinque e Seicento* (Soveria Mannelli: Rubbettino, 2001), 246–54.

27. Parts 2 and 4 of a four-volume work, *Doctrina Sacri Concilij Trid. et Catechismi Romani de sacramentis, de iustificatione, in symbolum Apostolorum, & in Decalogum, fideliter collecta, distincta, & vbi opus est, explicata* (Brescia: 1610–11)

28. This had been most recently printed in Venice in 1601 by Roberto Meietti.

29. See Dennis E. Rhodes, *Silent Printers: Anonymous Printing at Venice in the Sixteenth Century* (London: British Library, 1995), for an important analysis of false imprints using typographical evidence.

30. The "Religio" title page device is also used in a larger format for folio books: in Pulciano's 1607 *Pratica medica* of Alessandro Massaria, actually a reissue, with new title page, of a 1606 Treviso edition printed by Deuchino and funded by Meietti; in Marc'Antonio Pellegrini's *De fideicommissis praesertim vniuersalibus*, published in 1603 in Venice by Meietti and reissued in 1607 (as a new, seventh, edition) at the "Sign of Italy" ("sub Signo Italiae") with the "Religio" printer's mark replacing Meietti's mark on the new title page (the edition still contains the notice "Typographus Lectori" from the 1603 edition, signed by Meietti); in another work by Pellegrini, the six-volume *Consiliorum siue Responsorum*, a joint publication by Meietti and Deuchino (the third volume printed in 1606 by Roberto Meietti's father, Paolo, and reissued in 1607 with a new title page featuring the "Religio" sign, but with no publisher or printer); and in Giacomo Zabarella's *De rebus naturalibus, libri 30*, printed in Treviso by Deuchino with backing from Meietti in 1604 and reissued with the new "Religio" mark on the title page in 1608 in Venice, again at the "Sign of Italy" ("Sub signo Italiae"). The 1608 reissue is miscatalogued in the Servizio Bibliotecario Nazionale as a Bonifacio Ciera publication. For a full list of Zabarella imprints, offering much suggestive evidence for business contacts between Venice and Frankfurt, see Ian Maclean, "Mediations of Zabarella in Northern Germany, 1586–1623," in *Learning and the Market Place: Essays in the History of the Early Modern Book* (Leiden: Brill, 2009).

31. The one exception is the *Rotonda*, published in 1609, which does not carry the "Religio" mark. Instead, it revives the disused mark of Cornelio Arrivabene, who stopped printing in 1598. It is unclear why Baglioni switched marks for this one edition.

32. "Ecelente Signor, m'è stà ordenà / Da tutti quanti quei de Stamparia, / Che rengratia la Vostra Signoria / Del bel presente, che la n'ha donà. / Così preghemo Dio che come El g'ha / Dà gratiar in reprovar chi l'ha tradia / A torto, che in favor sempre 'l ghe sia / Per tutto 'l mondo e dove El l'ha esaltà: / E qual volta ghe accada de stampar / Qualch'altro parto del so bel inzegno, / Che impedissa a le Capre el rampegar. / El nostro humil servir ghe demo in pegno, / Quando però la se vorrà degnar / De comandarne, e no passar sto segno." *OG* 29:576. The sonnet, now in BNCF, Ms. Gal. 316, f. 358, is signed "De la V[ostra] Sig[nori]a Ec[cellentissim]a / Devoti Ser[vito]ri / I stamp[ato]ri de la so opera, / E Bonif[ici]o in nome de tutti de bottega."

33. Copy consulted: Biblioteca Casanatense, Editti e bandi, Per. Est. 18.4 (f. 77b). The edict was also printed in Bologna by the Benetiana press with the same text. Reproduced from the copy at Archivio Arcivescovile di Bologna, Miscellanee vecchie, n. 774, f. 200 by Gian Luigi Betti, "L'interdetto di Venezia e Bologna," in *Ripensando Paolo Sarpi*, ed. Corrado Pin (Venice: Ateneo Veneto, 2006), 301. It was translated into English in William Harris Rule, *History of the Inquisition*, vol. 2 (New York: Scribner, 1874), 311–12.

34. P. F. Grendler, "Books for Sarpi: The smuggling of prohibited books into Venice during the interdict of 1606–1607," in *Essays Presented to Myron P. Gilmore* (Florence: La Nuova Italia, 1978), 1:109n17.

35. See de Vivo, *Information and Communication in Venice*, 221n123; and Rhodes, *Silent Printers*, 173–74. There were also other books that were planned, but which either were never printed or did not survive, such as an anti-Jesuit volume by Sarpi that Meietti was supposed to publish in 1607. Lazzerini, "Officina sarpiana," 29, citing Frajese, *Sarpi Scettico*, 213.

36. Gian Camillo Glorioso, *Responsio Ioannis Camilli Gloriosi ad vindicias Bartholomaei Soueri; item Responsio eiusdem ad scholium Fortunii Liceti* (Naples: 1630), 4. Camillo's early bi-

ographer, Tomasini, claimed that Camillo liked to spend whole days in Venetian bookshops engaging strangers in conversation: Jacopi Philippi Tomasini, *Elogia Virorum Literis* (Padua: 1644), 312; cited in Favaro, *Galileo Galilei e lo Studio di Padova*, 2:15. See also De Vivo, *Patrizi, informatori, barbieri*, 221.

37. Lincei, 39 B 7 (formerly 163), ff. 414r–421v, "Avvisi d'un pio Religioso per le Cose di Venezia" (anonymous, undated, but addressed to "Cardinale Giustiniano," presumably Benedetto Giustiniani, papal legate to Bologna in 1606: see entry in *DBI*). The account of Marsilio is on f. 417v–418r.

38. Paul F. Grendler, *The Roman Inquisition and the Venetian Press*, 1540–1605 (Princeton, NJ: Princeton University Press, 1977), 280.

39. On Belluzzi, also known as Bellucci and Sanmarino, see the entry in *DBI* and Horst de la Croix, "The Literature on Fortification in Renaissance Italy," *Technology and Culture* 4 (1963).

40. For Belluzzi, see the 1598 Michelmas Frankfurt Book Fair catalog, D3v. For Lanteri, see 1601 Michelmas, F3v. Both works are listed under the heading "vernacular" rather than by discipline. For these and other book fair catalogues, I used the digital edition published by Olms in the collection *Michaelismesse 1594–Michaelismesse 1691* (Leipzig: Verlegt von Grosse/Lamberg/Latomus, 1594–1691), originally edited by Bernhard Fabian and published on microfiche, accessed from the Wellcome Library, London.

41. Roberto Meietti, *Catalogus eorum librorum omnium, qui in ultramontanis regionibus impressi apud Roberti Meietti ad insignia duorum Gallorum* (Venice: 1602). Copy consulted: Bibliotheca Angelica, Rome: ZZ.21.34. Colophon: "Il Fine del Catalogo della Fiera di Pasqua 1602. Fatta da Tommaso Baglioni." For a discussion of these publishers' catalogs, see Alfredo Serrai, *Cataloghi (Tipografici. Editorali. Di librai. Bibliotecari)*, vol. 4 of *Storia della Bibliografia* (Rome: Bulzoni, 1993), 34. Ciotti published a similar catalog the same year: *Catalogus eorum librorum omnium, qui in ultramontanis regionibus impressi apud Io. Baptistam Ciottam*. See Dennis E. Rhodes, "Spanish Books on Sale in the Venetian Bookshop of G. B. Ciotti, 1602," *The Library: The Transactions of the Bibliographical Society* 12 (2011); Dennis E. Rhodes, *Giovanni Battista Ciotti (1562–1627?): Publisher Extraordinary at Venice* (Venice: Marcanium, 2013), 73–75; Marina Venier, "Immagini e documenti," in *Il libro italiano del cinquecento: Produzione e commercio* (Rome: Istituto Poligrafico e Zecca dello Stato, 1989), 41.

42. Serrai notes that the Jesuit polemicist and bibliographer Antonio Possevino recommended the use of the 1602 catalogues of Meietti and Ciotti for revisions of his *Bibliotheca Selecta* (first edition 1593) and *Apparatus Sacer* (first edition 1603). See Serrai, *Cataloghi*, 33 and 720.

43. Grendler, "Books for Sarpi," 107–9.

44. Savio, "Per l'epistolario di Paolo Sarpi (cont.)," *Aevum* 16:9.

45. The Avosto pamphlet carries Meietti's printer's mark.

46. ASV, Riformatori dello studio di Padova, 168. Document reference kindly supplied by Mario Infelise and Filippo de Vivo. "Elenco di librai ai cui si notifica:[. . .] Tommaso Baglioni agente de ms Roberto Meietti alla bottega de doi galli."

47. *OG* 11:531.

48. Vincenzo Spampanato, "Nuovi Documenti intorno a Negozi e Processi dell'Inquisizione, 1603–1624 (cont.)," *Giornale critico della filosofia italiana* 5 (1924): 374–75.

49. *Lettere d'Uomini Illustri: Che fiorirono nel principio del secolo decimsettimo, non più stampate* (Venice: 1744), 201–2.

50. Ibid., 210.

51. ASV, Santo Ufficio, 77; reference generously supplied by Mario Infelise.

52. See Federico Barbierato, *Nella stanza dei circoli: Clavicula Salmonis e libri di magia a Venezia nei secoli XVII e XVIII* (Milan: Sylvestre Bonnard, 2002) 183 and 211–18.

53. See Mario Infelise, "Ricerche sulla fortuna editoriale di Paolo Sarpi (1619–1799)," in *Ripensando Paolo Sarpi*, ed. Corrado Pin.

54. Favaro, *Galileo Galilei e lo Studio di Padova*, 2:135. The note mentions only four copies of Gualterotti's "poemi" lent to Bolzetta, and two to Meietti. It is possible that Paolo Meietti in Padua is meant here, rather than Roberto in Venice, as Bolzetta was active in both cities at this time. Galileo also lent or loaned (possibly for sale or reprint) a copy of his father Vincenzo's *Dialogo*.

55. Gaetano Cozzi, *Paolo Sarpi tra Venezia e l'Europa*.

56. The publisher is named merely as "Apud Societatem Venetam," following the details of the previous entry, for Guidobaldo Dal Monte's *Problematum astronomicorum libri septem* (Venice: Giunti, Ciotti "& socios," 1609). The entry reproduces, with slight differences, the wording of the title page and omits the final phrase that makes the moons of Jupiter the property of the Medici. This omission may be due to either the secrecy or the incertitude of the dedication.

57. Information supplied by Paul Needham. In 1602, Van Roomen wrote to Bologna from the Frankfurt fair on 10 April. Easter was 7 April. Favaro, ed., *Carteggio*, 250. Ottavio Cotogna's *Compendio delle poste* (Milan: 1623), book 5, p. 506, claims that the Frankfurt fair starts fifteen days before Easter Sunday and lasts five days. Cotogna may have been wrong.

58. Kepler reveals in the printed postscript to his *Dissertatio cum Nuncio Sidereo* (Prague, 1610) [35] that the spring catalogue had reached Prague by 20 April 1610. It is hard to imagine this stage taking less than a week, so the catalogue was almost certainly in print before the end of the fair.

59. Pierre Sardella, "Nouvelles et speculations à Venise au debut du XVIe siècle," *Cahier des Annales* 1 (1948): 57.

60. Magini sent just such a sheet to publicize his forthcoming *Primum Mobile* in 1603 to his correspondent van Roomen. See Favaro, ed., *Carteggio*, 439 (10 March 1603).

61. Needham, *Galileo Makes a Book,* 198.

62. See *GO* 10:366 (31 May 1610), where Hasdale apologizes for asking Galileo for a copy, "at that time not knowing that they had been sent to Frankfurt."

63. "Delaying and deferring publication [*publicazione*] was too dangerous and risky for me, as someone else might have anticipated me." *OG* 10:298 (Galileo to Vinta, draft, 19 March 1610, Padua).

64. *OG* 3:105 and 10:320 (19 April 1610).

65. Favaro, ed., *Carteggio*, 447. Magini to Cav. Aderbale Minerbio, Bologna, 29 December 1608.

66. *OG* 12:49, Cesi to Galileo, 12 April 1614.

67. The only other Venetian printer with strong enough ties to the fair was Ciotti—who was, however, in Sicily on a business trip when Galileo discovered the moons of Jupiter, was shipwrecked, and did not return until after the publication of the *Sidereus*. See Sarpi, *Lettere ai Protestanti*, 1:70–71, and P. Savio, "Per l'epistolario di Paolo Sarpi (cont.)," *Aevum* 14 (1940): 3–12. In general, see Rhodes, *Giovanni Battista Ciotti*.

68. See Lucien Febvre and Henri-Jean Martin, *The Coming of the Book: The Impact of Printing 1450–1800* (London: Verso, 1976), 228–33; Angela Nuovo, *The Book Trade in the Ital-*

ian Renaissance (Leiden: Brill, 2013), 281–95; Ian Maclean, *Scholarship, Commerce, Religion: The Learned Book in the Age of Confessions, 1560–1630* (Cambridge, MA: Harvard University Press, 2012); Reeves, *Galileo's Glassworks*, 150.

69. See Lazzerini, "Officina Sarpiana," 29–80.

70. One "Iean Petit" did publish a French book against tyrannicide in 1610, but he was based in Rouen: [M. Roussel], *Antimariana ou refutation des propositions de Mariana: Pour monstrer que la vie des Princes souverains doit ester inviolable aux subjects, & à la Republique* (Rouen: Iean Petit, 1610). The *Anti-Cotone* bears no typographical resemblance to this book.

71. R. A. Sayce, "Compositional Practices and the Localization of Printed Books, 1530–1800," *The Library* 21 (1966).

72. For Sarpi's view of post-Interdict censorship, see Savio, "Per l'epistolario di Paolo Sarpi (cont.)," 13 (1939): 601. See also Sarpi, *Lettere ai Protestanti*, 1:54 (letter to Francesco Castrino, 13 October 1609), for the description of a plan by Ciotti to import or reprint works such as De Thou's *Historia*, the second part of which was placed on the Index a month later.

73. For Sarpi on the *Anti-Cotone*, see Savio, "Per l'epistolario di Paolo Sarpi (cont.)," 14 (1940): 63. For the comments of the papal nuncio to cardinal Borghese, see ibid., 70n13 (4 December 1610). See also the remarks by Giovanni Mocenigo, the Venetian ambassador to Rome, to the doge and Senate on Paul V's knowledge of the book and its true publisher: "There has appeared a book called *Anti-Cotton*; it runs to about eighty pages and, with heretical licence, it speaks ill of the Catholic religion and of the Apostolic Chair. This book has been printed by Mejetti, in Venice, in Italian; in many copies there is the imprint of Lyons, but in fact it was printed in Venice, for both type and paper are recognized." *CSP, Ven.* 12:97, 18 December 1610.

74. Savio, "Per l'epistolario di Paolo Sarpi (cont.)," 14 (1940): 70.

75. "Veneti Inquisitoris lectae litterae affirmantis Celium Malaspinam Venetijs non degere, sed Mantuam profectum esse, et librum eius Novellarum Impressum fuisse à Meietto cum licentia ficta absque nomine impressoris." ACDF, Indice, Diari, II, f. 16r, (citation generously supplied by Mario Infelise). See also Elisa Rebellato, *La fabbrica dei divieti* (Milan: Bonnard, 2008), 69, which discusses the case without citing this document.

76. Guglielmo Enrico Saltini, "Di Celio Malespini ultimo novelliere italiano in prosa del secolo XVI," *Archivio Storico Italiano*, 5th ser., 13 (1894): 71–74, Doc. III, "Proposta che fa Celio Malespini alla Repubblica di Venezia con una petizione diretta a Sua Serenità il Doge e al eccelso Consiglio de' X"; original at ASV, Cons. X, Reg. Sec. 12, f.15.

77. Their critical apparatus was supplied by Guido Casoni, who then became a client of the Sagredo family, including them in his *Ode*.

78. Saltini, "Di Celio Malespini," 73.

79. On Santini, see Biagioli, *Galileo's Instruments of Credit*, 86–87 and 90–91; and Eileen Reeves, "Variable Stars: A Decade of Historiography on the *Sidereus Nuncius*," *Galilaeana* 8 (2011): 43–44.

Chapter Seven

1. Bucciantini, Camerota and Giudice, *Il telescopio*, chapter 3. It is not at all clear whether they are right, and it does not much matter; only three satellites are shown, but this does not necessarily imply that the observation must be dated before 13 January, when the fourth moon was spotted. It remained extremely common for Galileo to see only three or even two satellites for the entire period of his *Sidereus* observations. Moreover, on the back of the Aleppo letter

Jupiter is depicted as a circle, whereas in the two other extant manuscripts recording the first week's observations (now in the National Library in Florence, Ms. Gal. 48, and in Ann Arbor, Michigan) it is drawn as a spoked wheel. In the Ann Arbor sheet the period 7–15 January 1610 is covered; in the Florentine manuscript, the so-called "observation notes," the wheel changes to a circle during the night of the 15 January. The only other trace we possess of this crucial first week's observation is a facsimile reproduction of a schematic representation of a photograph of a now lost letter, dated 7 January 1610 and formerly kept in the Vatican, written by Galileo to an unidentified recipient (usually considered to be Antonio de' Medici or Enea Piccolomini). In that reproduction, thus removed from the autograph mark, Jupiter is represented as a large six-pointed asterisk; we lack evidence, however, to know whether this faithfully represents Galileo's sign or is Favaro's editorial interpretation, standardizing the apparently unimportant distinction between spoked wheel, circle, and asterisk to the norm adopted by the printers of the *Sidereus nuncius*.

2. After the first page of notes Galileo switches from Italian to Latin, and this is usually interpreted as the moment when he realized he would be communicating with an international audience and had a discovery really worth reporting. Above the first Latin observations, however, is this curious note in Italian: "Have them cut in wood all of them in one piece, the stars white, the rest black, then saw it into pieces" (*faransi i[n]tagliar i[n] legno tutte in u[n] pezzo, et le stelle bia[n]che il resto nero poi si seghera[n]no i pezzi*). BNCF, Mss. Gal. 48, f.30v.

3. Owen Gingerich and Albert Van Helden, "From 'Occhiale' to printed page: The making of Galileo's 'Sidereus Nuncius,'" *Journal for the History of Astronomy* 34 (2003).

4. The term used for a frisket was "fraschetta," but "maschera" appears as a term for stopping out in etching. Galileo's word for the telescopic diaphragm is lacking: he referred only to the hole, which in Italian he called "pertuso" and in Latin "foramen." "Foramen" was the standard term for the aperture of a camera obscura. The equivalence may echo Dante, *Inferno* 33:22–27, which uses both "pertugio" and "forame" to describe Count Ugolino's pinhole window, through which he glimpsed the moon in the prelude to his infamous fate: "Breve pertugio dentro da la Muda, / la qual per me ha 'l titol de la fame, / e che conviene ancor ch'altrui si chiuda, / m'avea mostrato per lo suo forame / più lune già, quand' io feci 'l mal sonno / che del futuro mi squarciò 'l velame" (A tiny chink inside that tower, which, because of me, is called Famine, and in which others will still be enclosed, had already shown me through its opening several moons, when I slept an evil sleep that ripped the curtain of the future for me). For details of the telescopic diaphragm, see Sven Dupré, "Galileo's Telescope and Celestial Light," *Journal for the History of Astronomy* 34 (2003). For the suggestion that the shape of the diaphragm was oval rather than circular because of Sarpi's anatomisation of cat and lynx eyes, see my "The Strangest Piece of News" (review of *Galileo: Watcher of the Skies* by David Wootton and *Galileo* by John Heilbron), *London Review of Books*, 2 June 2011, p. 31.

5. In fact, the humor of Wotton's English witticism was lost in his Latin translation: "Legatus est vir bonus missus ad meniendum Reip. Cause." The line later provoked a scandal, with an attack by Scioppius that Wotton refuted with a published letter to Welser. See Henry Wotton to Welser, London, 2 December 1612, in Smith, *Life and Letters*, 2:9. The letter was translated into English in the third edition of the *Reliquiae Wottonianae* (London: Printed by T. Roycroft for R. Marriott, F. Tyton, T. Collins, and J. Ford, 1672; first edition 1651).

6. See my "Galilean Angels," in *Conversations with Angels: Essays Towards a History of Spiritual Communication, 1100–1700*, ed. Joad Raymond (New York: Palgrave Macmillan, 2011).

7. *OG* 10:342–43, quoted in Kepler, *Dissertatio*, ed. Pantin, xxv; partially quoted in Heilbron, *Galileo*, 163–64, translation mine.

8. See the classic case of false identity in Natalie Zemon Davis, *The Return of Martin Guerre* (Cambridge, MA: Harvard University Press, 1984) and the more general enquiry of Valentin Groebner, *Identification, Deception, and Surveillance in Early Modern Europe* (Cambridge, MA: MIT Press, 2007).

9. On Ṣafar, see Rota, "Safavid Envoys in Venice."

10. Sarpi, *Lettere ai Protestanti*, 1:56 (Sarpi to Francesco Castrino, 13 October 1609).

11. ASF, Mediceo del Principato, 302, f.57, 5 January 1610.

12. ASF, Mediceo del Principato, 302, f.59, 16 January 1610.

13. Angelo Michele Piemontese, "I due Ambasciatori di Persia ricevuti da papa Paolo V al Quirinale," *Miscellanea Bibliothecae Apostolicae Vaticanae* 12 (2005).

14. See, for example, the 1603 painting in the Sala delle Quattro Porte of the Palazzo Ducale, Venice, by Carlo and Gabriele Caliari, *Doge Marino Grimani Receiving the Persian Ambassadors*, and the other objects in the exhibition *Gifts by Shah Abbas the Great to the Serenissima: Diplomatic Relations between the Serenissima and Safavid Persia* at the Palazzo Ducale, Sala delle Quattro Porte, 28 September 2013 to 12 January 2014. This painting is discussed in depth in Rota, "Safavid Envoys in Venice."

15. Dragomans have been brilliantly studied by Nathalie Rothman, *Brokering Empire: Trans-Imperial Subjects between Venice and Istanbul* (Ithaca, NY: Cornell University Press, 2011).

16. Smith, *Life and Letters*, 1:414.

17. *Sbandimento, esamine e processo del fraudolente, insolente e prodigo Carnevale, con la rinuncia, ch'ei fa, avanti che faccia partenza di questi nostri paesi. Il quale è bandito per un anno, et secondo che parerà a' suoi maggiori,* (Bologna: 1624); *Tragedia in comedia fra i bocconi di grasso e quei di magro la sera di Carnevale. Con il lamento del Carnevale, dolendosi della Quaresima, che li sia sopragiunta così presto. Et la risposta di lei contro il Carnevale. Capriccio galante,* (Bologna: [s.d.]); *Processo overo esamine di Carnevale. Nel quale s'intendono tutti gl'inganni, astutie, capriccij, bizarie, viluppi, intrichi, inventioni, novità, sottilità, scioccharie, grillarie, etc., ch'egli ha fatto quest'anno nella nostra città. Con la sentenza et bando contro di lui formata.* (Bologna: 1588), etc. See *sub voce* in *DBI*.

18. *Lettere del molto R.P. abbate D. Angelo Grillo,* (Venice: 1608), imprimatur 20 June 1602, p. 635. Grillo writing to Angelo Mariani, Pavia, no date.

19. Jean-Pierre Cavaillé, *Dis/simulations: Jules-César Vanini, François La Mothe Le Vayer, Gabriel Naudé, Louis Machon et Torquato Accetto. Religion, morale et politique au XVIe siècle* (Paris: Champion, 2002); Jon Snyder, *Dissimulation and the Culture of Secrecy in Early Modern Europe* (Berkeley: University of California Press, 2012). For a study of Venice in the eighteenth century, see James Johnson, *Venice Incognito: Masks in the Serene Republic* (Berkeley: University of California Press, 2011).

20. An excellent point of access to the vast field of Sarpi scholarship is Corrado Pin's *DBI* entry. For Descartes, see Jean-Luc Nancy, *Ego sum* (Paris: Aubier-Flammarion, 1979); and Jean-Pierre Cavaillé, *Descartes, la fable du monde* (Paris: Vrin, 1991).

21. Ludovico delle Colombe, *Risposte piaceuoli, e curiose di Lodouico delle Colombe alle considerazioni di certa maschera saccente nominata Alimberto Mauri, fatte sopra alcuni luoghi del discorso del medesimo Lodouico dintorno alla stella apparita l'anno 1604* (Florence: 1607).

22. *Serenissimo et Pietoso aviso della Chiesa apostolica Romana al santissimo signor Paulo Quinto Pontifice Massimo è moderno Vescouo di Roma. Sopra la differenza tra la sua santità & la sereniβima signoria di Venetia. Alquale si dimostra con chiarissime leggi è ragioni diuine è di politia Che si l'apostolico Vuol menar la guerra Non sara piu apostolico sopra la terra* (Stampata nella Città Angelica al segno di S. Pietro & Paulo col priuilegio del Saluatore sempiterno, Anno

M. DC. VII.), kindly bought to my attention by Filippo de Vivo, and listed in De Vivo, *Patrizi, informatori, barbieri*, 398.

23. See Eileen Reeves and Albert Van Helden, *Galileo and Scheiner on Sunspots, 1611–1613* (Chicago: University of Chicago Press, 2010) for an introduction and full translations of the initial texts. In addition to the "original" debate, the following texts should be considered: Johannes Fabricius, *De maculis in Sole observatis et apparente earum cum Sole conversione narratio: Cui adiecta est de modo eductionis specierum visibilium dubitatio* (Wittenberg: 1611); Jean Tarde, *Borbonia Sidera* (Paris: 1621 [privilege dated 8 June 1620]), translated by the author as *Les Astres de Borbon* (Paris: 1622); Christoph Scheiner, *Rosa Ursina, sive sol ex admirando facularum & macularum suarum phoenomeno varius* (Bracciani: 1626–30); Galileo Galilei, *Dialogo di Galileo Galilei . . . sopra i due massimi Sistemi del mondo tolemaico, e copernicano* (Florence: 1632); and Carolus Malapertius, *Austriaca sidera heliocyclica astronomicis* (Douai: 1633).

24. See Biagioli, *Galileo, Courtier*.

25. Biagioli, *Galileo's Instruments*, 175.

26. On Welser, see Gabrielli, "Marco Welser Linceo," 74–99. More recent studies are listed in Reeves and Van Helden, *Galileo and Scheiner on Sunspots*.

27. Sagredo to Welser, 4 April 1614, *OG* 10:45.

28. For a description of the content and context of the *Squitinio*, see Bouwsma, *Venice*, 504.

29. *Squitinio della libertà veneta: Nel quale si adducono anche le raggioni dell'Impero Romano sopra la Città & Signoria di Venetia,* first printed with the false imprint of Mirandola in 1612 in two editions, and then again in 1619.

30. *Viri illustris Nicolai Claudii Fabricii de Peiresc senatoris aquisextiensis vita* (Paris: 1641).

31. [Jean] Baptiste Legrain, *Decade commencant l'Histoire du Roy Louys XIII* (Paris: 1618), 449. Welser is named, probably relying on diplomatic reports, as "Vulser." See also Henricus Ernestius, *Variarum Observationum Libri Duo* (Amsterdam: 1636), book 2, chapter 36, p. 160.

32. Vicentius Placcius, *Theatrum anonymorum et pseudonymorum ex symbolis & collatione vitorum per Europam doctissimorum ac celeberrimorum* (Hamburg: 1708).

33. Christoph Arnold, *Marci Velseri Opera in unum collecta* (Nuremberg: 1682), 3.

34. ASV, Inquisitori di Stato, Dispacci Ambasciatori da Vienna, Busta 1259 (5 November 1612).

35. Ibid.

36. ASV, Cons X (Secreta) 15, f.123r (19 November 1612), f.122r (16 November 1612).

37. ASV, CC.X. Lettere. Amb. 13 (Germania) ff. 229r–229av (8 December 1612, letter received 16 December).

38. Shelfmark—D 137D 127.

39. See Reeves and Van Helden, *On Sunspots*, 87–89.

40. *Disquisitione Mathematicae* (Ingolstadt, 1614).

41. Franciscus Aguillonius, *Opticorum Libri Sex* (Antwerp: 1613), book 5, propositio LVI, 421.

42. *OG* 10:51.

43. Steven Shapin, *A Social History of Truth: Civility and Science in Seventeenth-Century England* (Chicago: University of Chicago Press, 1994). Boyle's *Invitation to Free Communication* is reprinted in vol. 1 of *The Works of Robert Boyle,* ed. Michael Hunter and Edward B. Davis (London: Pickering & Chatto, 1999).

44. For a full bibliography see De Vivo, *Patrizi, informatori, barbieri*, 369–403.

45. Galileo's version of the debate runs as follows: [Anonymous, i.e. Orazio Grassi], *De tribus cometis anni MDCXIII Disputatio astronomica publice habita in Collegio Romano Societatis Iesu ab uno ex patribus eiusdem Societatis* (Rome: 1618), Mario Guiducci, *Discorso delle comete* (Florence, 1619); Lothario Sarsio [i.e., Orazio Grassi], *Libra astronomica ac philosophica, qua Galilaei Galilaei opiniones de comits a Mario Guiduccio in Florentina Academia expositae, atque in lucem nuper editae, examinantur a Lothario Sarsio Sigensano* (Perugia: 1619). This debate was supplemented by Mario Guiducci, *Lettera al M. R. P. Tarquinio Galluzzi della Compagnia di Giesù* (Florence: 1620).

46. The copy, with the inscription "Al S.r Cavalier Tommaso Stigliani l'Autore," was offered for sale by Philobiblon in 2008, and by Martayan Lan, and is now in a private collection in Virginia. It appears to be genuine.

47. *OG* 6:219. The best edition of *Il Saggiatore* is by Ottavio Besomi and Mario Helbing (Rome and Padua: Antenore, 2005).

48. *OG* 6:219–20.

49. *OG*, 6:114.

50. Ibid.

51. *OG*, 6:141.

52. Plato, *Epistle II* (to Dionysius), Loeb, IX, 403–19, 415–17. The letter referred to functions as one of the points of access to Jacques Derrida's *The Post-Card: From Socrates to Freud and Beyond* (Chicago: University of Chicago Press, 1987) where the shifting positions of Socrates and Plato and the self-dismantling hierarchies of speech and writing create a kind of *mise-en-abime* which undoes the "myth of presence" in Western metaphysics. Derrida's analysis seems pertinent here because it draws attention to narratives of power that accompany models of representation. The self-effacement of writing, which takes place in writing (in a letter which "should" not exist, and whose existence is a testament to betrayal by its reader) accompanies a crisis in authorship.

53. *OG*, 12:454.

54. *OG*, 12:460–61. On the *Assemblea Celeste* of Giacomo or Giovanni Rho, an anonymous Jesuit cometary tract published in Milan, see Ottavio Besomi and Michele Camerota, *Galileo e il Parnaso tychonico: Un capitolo inedito del dibattito sulle comete tra finzione letteraria e trattazione scientifica* (Florence: L. Olschki, 2000).

Bibliography

Aguillonius, Franciscus. *Opticorum Libri Sex.* Antwerp: 1613.

Alberi Auber, Paolo. "Una miniera, un forno per il ferro e due uomini di Scienza fra le montagne: Nicola Cusano e Gianfrancesco Sagredo." *Archivio per l'Alto Adige* 100 (2006): 1–100.

Alden, Dauril. *The Making of an Enterprise: The Society of Jesus in Portugal, Its Empire, and Beyond, 1540–1750.* Stanford, CA: Stanford University Press, 1996.

Arslan, Edoardo. *I Bassano.* Milan: Ceschina, 1960. 2 vols.

Baldini, Ugo. "The Jesuit College in Macao as a Meeting point of the European, Chinese and Japanese Mathematical Traditions: Some Remarks on the Present State of Research, Mainly Concerning Sources, 16th–17th centuries." In *The Jesuits, the Padroado and East Asian science, 1552–1773: History of mathematical sciences: Portugal and East Asia III*, edited by Luís Saraiva and Catherine Jami, 33–80. Singapore: World scientific, 2008.

———. *Legem impone subactis: Studi su filosofia e scienza dei Gesuiti in Italia, 1540–1632.* Rome: Bulzoni, 1992.

Baldini, Ugo, and Leen Spruit. *Catholic Church and Modern Science: Documents from the Archives of the Roman Congregations of the Holy Office and the Index.* 4 vols. Rome: Libreria Editrice Vaticana, 2009.

Bando che nessuno possa scriver Lettere d'Avisi, Senza Licenza di Monsignor Governatore. Rome, Nella Stamperia della Camera Apostolica, 19 September 1602.

Barbierato, Federico. *Nella stanza dei circoli: Clavicula Salmonis e libri di magia a Venezia nei secoli XVII e XVIII.* Milan: Sylvestre Bonnard, 2002.

Barlow, William. *Magnetical Advertisements Concerning the Nature and Property of the Loadstone.* London: 1616.

Batchelor, Robert K. London: *The Selden Map and the Making of a Global City, 1549–1689.* Chicago: University of Chicago Press, 2014.

Bellarino, Giovanni. *Doctrina Catechismi Romani.* Vols. 2 and 4 of *Doctrina Sacri Concilij Trid. et Catechismi Romani de sacramentis, de iustificatione, in symbolum Apostolorum, & in Decalogum, fideliter collecta, distincta, & vbi opus est, explicata.* Brescia: 1610–11.

Ben-Zaken, Avner. *Cross-Cultural Scientific Exchanges in the Eastern Mediterranean, 1560–1660.* Baltimore: Johns Hopkins University Press, 2010.

Berchet, Guglielmo. *Relazioni dei Consoli Veneti nella Siria.* Turin: Paravia, 1866.

Berenson, Bernard. *Italian Pictures of the Renaissance: A List of the Principal Artists and Their Works, with an Index of Places: Venetian School.* London: Phaidon Press, 1957.

Bertoloni Meli, Domenico. *Thinking with Objects: The Transformation of Mechanics in the Seventeenth Century.* Baltimore: Johns Hopkins Press, 2006.

Besomi, Ottavio, and Michele Camerota. *Galileo e il Parnaso tychonico: Un capitolo inedito del dibattito sulle comete tra finzione letteraria e trattazione scientifica.* Florence: L. Olschki, 2000.

Betti, Gian Luigi. "L'interdetto di Venezia e Bologna." In *Ripensando Paolo Sarpi*, ed. Corrado Pin, 271–306. Venice: Ateneo Veneto, 2006.

Biagioli, Mario. "Did Galileo Copy the Telescope? A 'New' Letter by Paolo Sarpi." In *The Origins of the Telescope*, ed. Albert Van Helden, Sven Dupré, Rob van Gent, and Huib Zuidervaart, 203–30. Amsterdam: KNAW, 2010.

———. *Galileo Courtier: The Practice of Science in the Culture of Absolutism.* Chicago: University of Chicago Press, 1993.

———. *Galileo's Instruments of Credit.* Chicago: University of Chicago Press, 2006.

Binion, Alice. "Algarotti's Sagredo Inventory," *Master Drawings* 21 (1983): 392–96.

Bolt, Marvin, and Michael Korey. "Trumpeting the Tube: A Survey of Early Trumpet-Shaped Telescopes." In *Der Meister und die Fernrohre: Das Wechselspiel zwischen Astronomie und Optik in der Geschichte*, edited by Jürgen Hamel, Rolf Riekher, and Inge Keil, 146–63. Frankfurt: Harri Deutsch Verlag, 2007.

Borean, Linda. " 'In camera dove dormo': Su alcuni quadri di Nicolò Sagredo," *Arte veneta* 50 (1997): 122–30.

Bouwsma, William J. "The Renaissance and the Drama of Western History," *American Historical Review* 84 (1979): 1–15.

Bouwsma, William J. *Venice and the Defense of Republican Liberty: Renaissance Values in the Age of the Counter Reformation.* Berkeley and Los Angeles: University of California Press, 1968.

Boyle, Robert. *Invitation to Free Communication.* In vol. 1 of *The Works of Robert Boyle*, edited by Michael Hunter and Edward B. Davis. London: Pickering & Chatto, 1999.

Brahe, Tycho. *Tychonis Brahe Dani Opera Omnia.* Edited by Johan Ludwig Emil Dreyer and Hans Raeder, 15 vols. Amsterdam: Swets and Zeitlinger, 1972.

Bray, Alan. *The Friend.* Chicago: University of Chicago Press, 2003.

Bredekamp, Horst. *Galilei der Künstler: Der Mond. Die Sonne. Die Hand.* Berlin: Akademie Verlag, 2007.

Bredekamp, Horst, ed. *Galileo's O.* 2 vols. Berlin: Akademie Verlag, 2011.

Bucciantini, Massimo. *Galileo e Keplero: Filosofia, cosmologia e teologia nell'Età della Controriforma.* Turin: Einaudi, 2003.

Bucciantini, Massimo, Michele Camerota, and Franco Giudice. *Il Telescopio di Galileo: Una storia europea.* Turin: Einaudi, 2012.

Burggrav, Joann Ernest. *Achilles Πανοπλος Redivivus, seu Panoplia Physico-Vulcania quâ in praelio φιλοπλος in Hostem educitur Sacer et inviolabilis.* [Amsterdam]: 1612.

Calepino, Ambrogio. *Dictionarium in quinque linguis.* Venice: 1590.

Camerota, Filippo. *Il Compasso di Fabrizio Mordente: Per la storia del compasso di proporzione.* Florence: Leo S. Olschki, 2000.

Camerota, Michele. *Galileo Galilei e la cultura scientifica nell'Età della Controriforma,.* Rome: Salerno, 2004.

Cappelletti, Giuseppe. *I Gesuiti e la Republica di Venetia.* Venice: Grimaldo, 1873.

Carli, Alarico and Antonio Favaro, eds., *Bibliografia galileiana 1565–1895*. Rome: Ministero della Pubblica Istruzione, 1896.

Carolino, Luís Miguel, and Carlos Ziller Camenietzki, eds., *Jesuítas, ensino e ciência, séc. XVI–XVIII*. Casal de Cambra: Caleidoscópio, 2005.

Carugo, Adriano, and Alistair C. Crombie. "The Jesuits and Galileo's Ideas of Science and of Nature." *Nuncius* 8 (1983): 3–68.

Casoni, Guido. *Ode dell'illust. et eccell. signore Guido Casoni dedicate all'illustriss. & reuerendiss. sig. cardinale Cinthio Aldobrandini*. Venice: Gio. Battista Ciotti, 1601 (2nd edn.).

———. *Vita del glorioso santo Gerardo Sagredo nobile venetiano, monaco dell'ordine di san Benedetto*. Venice: 1598.

Catalogue of Manuscripts in the British Museum, New Series, vol. I, part 2. *The Burney Manuscripts*. London: British Museum, 1840.

Cavaillé, Jean-Pierre. *Descartes, la fable du monde*. Paris: Vrin, 1991.

———. *Dis/simulations: Jules-César Vanini, François La Mothe Le Vayer, Gabriel Naudé, Louis Machon et Torquato Accetto. Religion, morale et politique au XVIe siècle*. Paris: Champion, 2002.

Cecchetti, Bartolomeo. *La republica di Venezia e la corte di Roma nei rapporti della religione*. Venice: Naratovich, 1874.

Chambers, David. "Merit and Money: The Procurators of St. Mark and their *Commissioni* 1443–1605," *Journal of the Warburg and Courtauld Institutes* 60 (1997): 23 88.

Cicogna, Emmanuele Antonio. *Delle inscrizioni veneziane*. 6 vols. Venice: Giuseppe Picotti, 1824–53.

Ciotti, Giovanni Battista. *Catalogus eorum librorum omnium, qui in ultramontanis regionibus impressi apud Io. Baptistam Ciottam*. Venice: 1602.

Clark, Stuart. *Vanities of the Eye*. Oxford: Oxford University Press, 2007.

Cole, Janie. "Cultural Clientelism and Brokerage Networks in Early Modern Florence and Rome: New Correspondence between the Barberini and Michelangelo Buonarroti the Younger." *Renaissance Quarterly* 60 (2007): 729–88.

Cozzi, Gaetano. *Il doge Niccolò Contarini: Ricerche sul patriziato veneziano agli inizi dei Seicento*. Venice and Rome: Istituto per la Collaborazione Culturale, 1958.

———. "Giuspatronato del doge e prerogative del primicerio sulla Cappella Ducale di San Marco (secoli XVI–XVIII): Controversie con i procuratori di San Marco de supra e i patriarchi di Venezia." In *Atti dell'Istituto veneto di scienze lettere ed arti* 151 (1992–93): 1–69.

———. "Note su Giovanni Tiepolo, primicerio di San Marco e patriarca di venezia: L'unità ideale della chiesa veneta." In *Chiesa, Società e Stato a Venezia: Miscellanea di studi in onore di Silvio Tramontin nel suo 75 anno di età*. Edited by Bruno Bertoli, 121–50. Venice: Studium Cattolico Veneziano, 1994.

———. *Paolo Sarpi tra Venezia e l'Europa*. Turin: Einaudi, 1978.

———. "Sulla morte di fra Paolo Sarpi," in *Miscellanea in onore di Roberto Cessi*, II: 387–96. Rome: Edizioni di storia e letteratura, 1958.

———. "Una vicenda della Venezia barocca: Marco Trevisan e la sua eroica amicizia," in *Venezia barocca—Conflitti di uomini e idee nella crisi del Seicento veneziano*. Venice: Il Cardo, 1995. Originally published in *Bollettino dell'Istituto di Storia della Società e dello Stato, Fondazione Giorgio Cini* 2 (1960).

Cremonini, Cesare. "*Lecturae Exordium*" in *Le Orazioni* ed. Antonio Poppi, 3–51. Padua: Antenore, 1998. Originally published in Ferrara, 1591.

Croce, Giulio Cesare. *Processo overo esamine di Carnevale. Nel quale s'intendono tutti gl'inganni, astutie, capriccij, bizarie, viluppi, intrichi, inventioni, novità, sottilità, scioccharie, grillarie, etc., ch'egli ha fatto quest'anno nella nostra città. Con la sententia et bando contro di lui formata.* Bologna: 1588.

———. *Sbandimento, esamine e processo del fraudolente, insolente e prodigo Carnevale, con la rinuncia, ch'ei fa, avanti che faccia partenza di questi nostri paesi. Il quale è bandito per un anno, et secondo che parerà a' suoi maggiori.* Bologna 1624.

———. *Tragedia in comedia fra i bocconi di grasso e quei di magro la sera di Carnevale. Con il lamento del Carnevale, dolendosi della Quaresima, che li sia sopragiunta così presto. Et la risposta di lei contro il Carnevale. Capriccio galante.* Bologna: [s.d.]

De la Croix, Horst. "The Literature on Fortification in Renaissance Italy." *Technology and Culture* 4 (1963): 30–50.

Della Porta, Giovan Battista. *Magiae naturalis libri XX.* Naples: 1589.

Delle Colombe, Ludovico. *Risposte piaceuoli, e curiose di Lodouico delle Colombe alle considerazioni di certa maschera saccente nominata Alimberto Mauri, fatte sopra alcuni luoghi del discorso del medesimo Lodouico dintorno alla stella apparita l'anno 1604.* Florence: 1607.

De Vivo, Filippo. "How to read Venetian *Relazioni.*" *Renaissance and Reformation* 34 (2011): 25–59.

———. *Information and Communication in Venice: Rethinking Early Modern Politics.* Oxford: Oxford University Press, 2007.

De Vivo, Filippo. *Patrizi, informatori, barbieri: Politica e comunicazione a Venezia.* Milan: Feltrinelli, 2012.

———. "Paolo Sarpi and the Uses of Information in Seventeenth-century Venice." In *News Networks In Seventeenth Century Britain And Europe* ed. Joad Raymond, 35–50. London: Routledge, 2005.

Derrida, Jacques. *The Post-Card: From Socrates to Freud and Beyond.* Chicago: University of Chicago Press, 1987.

Dondi, Raffaele. "Di Leone Bonzio incisor pubblico di anatomia a Venezia e del suo ritratto dipinto da Leandro da Ponte detto il Cavalier Bassano," *Rivista di Storia della Medicina* 4 (1960): 205–16.

Donne, John. *The Epigrams, Epithalamions, Epitaphs, Inscriptions, and Miscellaneous Poems,* edited by Ted-Larry Pebworth, Gary A. Stringer, Ernest W. Sullivan II, and William A. McClung. Volume 8 of *The Variorum Edition of the Poetry of John Donne,* edited by Gary Stringer. Bloomington: Indiana University Press, 1995.

Donnelly, John Patrick. "The Jesuit College at Padua: Growth, Suppression, Attempts at Restoration, 1552–1606." *Archivum Historicum Societatis Iesu* 51 (1982): 45–78.

Drake, Stillman. *Operations of the Geometric and Military Compass, 1606.* Washington: Dibner Library of the History of Science and Technology, 1978.

———. "The Starry Messenger." *Isis* 49 (1958): 346–47.

———. "Tartaglia's Squadra and Galileo's Compasso." *Annali dell'Istituto e Museo di storia della scienza di Firenze* 2 (1977): 35–54.

Dupré, Sven. "Galileo's Telescope and Celestial Light." *Journal for the History of Astronomy* 34 (2003): 369–99.

Edictum illustrissimorum & reuerendissimorum dominorum cardinalium generalium inquisitorum. Rome: 1606. Incipit: "Cum Robertus Meiettus librorum impressor Venetijs ausus sit . . ."

Ernestius, Henricus. *Variarum Observationum Libri Duo.* Amsterdam: 1636.

Evans, Julianne. "On the Character of Sagredo: An English Translation of Galileo's Judgments upon His Nativity of Giovanni Sagredo." *Culture and Cosmos* 7 (2003): 97–103.

Fabricius, Johannes. *De maculis in sole observatis et apparente earum cum sole conversione narratio: Cui adiecta est de modo eductionis specierum visibilium dubitatio.* Wittenberg: 1611.

Favaro, Antonio. *Amici e corrispondenti di Galileo.* Edited by Paolo Galluzzi, 3 vols. Florence: Salimbeni, 1983.

———. "Ancora a proposito di Giovanfrancesco Sagredo," in *Scampoli galileiani,* edited by Lucia Rossetti and Maria Laura Soppelsa, vol. 2, 505–10. Trieste: Lint, 1992. Originally published in *Atti e memorie della R. Accademia di scienze, lettere ed arti in Padova* 22 (1906): 9–14.

———. "Appendice seconda alla Libreria di Galileo." In *Scampoli galileiani,* edited by Lucia Rossetti and Maria Laura Soppelsa, 2 vols. Trieste: Lint, 1992. Originally published in *Atti e memorie della R. Accademia di scienze, lettere ed arti in Padova* 12 (1896): 44–50.

———. *Galileo Galilei e lo studio di Padova.* Padua: Antenore, 1966. First edition Florence, 1883.

———. "Giovanfrancesco Sagredo," in *Amici e corrispondenti di Galileo,* edited by Paolo Galluzzi, Florence, Libreria editrice Salimbeni, 1983): 191–322. Originally published in *Nuovo Archivio Veneto* 4 (1902): 313–422.

———. "Giovanfrancesco Sagredo e Guglielmo Gilbert" in *Adversaria galilaeiana: Serie I–VII,* edited by Lucia Rossetti and Maria Laura Soppelsa, 100–103. Trieste: Lint, 1992. Originally published in *Atti e memorie della R. Accademia di scienze, lettere ed arti in Padova* 35 (1919): 12–15.

———. "Giovanfrancesco Sagredo e la Vita Scientifica in Venezia al principio del XVII secolo." *Nuovo archivio veneto* 4 (1902): 313–442.

———. "Inventario della eredità di Galileo," in *Scampoli Galileiani,* edited by Lucia Rossetti and Maria Laura Soppelsa, vol. 1, 64–69. Trieste: Lint, 1992. Originally published in *Atti e memorie della R. Accademia di scienze, lettere ed arti in Padova* 4 (1888): 122–27.

———. "La libreria di Galileo Galilei descritta ed illustrata." *Bullettino di Bibliografia e di Storia delle Scienze Matematiche e Fisiche* 19 (1886): 219–93.

———. "Nuovi documenti sulla vertenza tra lo Studio di Padova e la compagnia di Gesù sul finire del secolo decimosesto." *Nuovo Archivio Veneto* 21 (1911): 89–100.

———. "Riccardo Willoughby." In *Amici e corrispondenti di Galileo.* Edited by Paolo Galluzzi, vol. 2, 1001–5. Florence: Salimbeni, 1983. Originally published in *Atti dell' I.R. Istituto veneto di scienze, lettere ed arti* 71 (1911/12): 25–29.

———. *Scampoli galileiani,* edited by Lucia Rossetti and Maria Laura Soppelsa, 2 vols. Trieste: Lint, 1992.

———. "Studi e ricerche per una iconografia galileiana," *Atti del Reale Istituto Veneto di Scienze, Lettere ed Arti* 72 (1913): 995–1051; 73 (1914): 105–34; 74 (1914): 305–13; and 75 (1915): 55–64.

———. "Lo Studio di Padova e la Compagnia di Gesù sul finire del secolo decimosesto." *Atti del Reala Istituto Veneto di Scienze, Lettere ed Arti* 4 (1877–78): 401–535.

Favaro, Antonio, ed. *Carteggio Inedito di Ticone Brahe, Giovanni Keplero e di altri celebri astronomi e matematici dei secoli XVI e XVII con Giovanni Magini.* Bologna: Zanichelli, 1886.

Febvre, Lucien, and Henri-Jean Martin. *The Coming of the Book: The Impact of Printing 1450–1800.* London: Verso, 1976.

Feingold, Mordechai, ed. *Jesuit Science and the Republic of Letters*. Cambridge, MA: MIT Press, 2003.

———. *The New Science and Jesuit Science: Seventeenth Century Perspectives*. Boston: Kluwer Academic Publishers, 2003.

Flayder, Friedrich Hermann. *De Arte Volandi*. Tubingen: 1627.

Fortini Brown, Patricia. *Private Lives in Renaissance Venice: Art, Architecture, and the Family*. New Haven: Yale University Press, 2004.

Foscarini, Marco. *Della letteratura veneziana ed altri scritti intorno ad essa*. Bologna: Forni, 1976; reprint of 1854 edn. (1st edn. 1752).

Frajesi, Vittorio. *Sarpi Scettico: Stato e Chiesa a Venezia tra Cinque e Seicento*. Bologna: Il Mulino, 1994.

Fraser, Peter Marshall. *Ptolemaic Alexandria*. 3 vols. Oxford: Clarendon Press, 1972.

Gabrieli, Giuseppe. "Marco Welser Linceo Augustano." *Rendiconti della R. Accademia Nazionale dei Lincei: Classe di scienze morali, storiche e filologiche* 13 (1937).

Galilei, Galileo. *Dialogo di Galileo Galilei . . . sopra i due massimi Sistemi del mondo tolemaico, e copernicano*. Florence: 1632.

———. *Dialogo sopra i due massimi sistemi del mondo tolomaico e copernicano*. Edited by Ottavio Besomi and Mario Helbing. 2 vols. Padua: Antenore, 1998.

———. *Difesa di Galileo Galilei [. . .] contro alle calunnie & imposture di Baldassar Capra Milanese*. Venice: 1607.

———. *Discorso [. . .] intorno alle cose che stanno in sù l'acqua ò che in quella si muovono*. Florence: 1612.

———. *Le Operazioni del compasso geometrico e militare*. Padua: 1606.

———. *Il Saggiatore*. Edited by Ottavio Besomi and Mario Helbing. Rome and Padua: Antenore, 2005.

———. *Sidereus nuncius*. Venice: 1610.

Garzoni, Leonardo. *Trattati della calamita*. Edited by Monica Ugaglia. Milan: FrancoAngeli, 2005.

Gassendi, Pierre. *Viri illustris Nicolai Claudii Fabricii de Peiresc senatoris aquisextiensis vita*. Paris: 1641.

Gerola, Giuseppe. *Bassano*. Bergamo: Istituto italiano d'arti grafiche, 1910.

Ghelfucci, Capoleone. *Il Rosario della Madonna*. Venice: 1600.

Gingerich, Owen, and Albert Van Helden. "From 'Occhiale' to Printed Page: The Making of Galileo's 'Sidereus Nuncius.'" *Journal for the History of Astronomy* 34 (2003): 251–67.

Glorioso, Gian Camillo. *Responsio Ioannis Camilli Gloriosi ad vindicias Bartholomaei Soueri; item Responsio eiusdem ad scholium Fortunii Liceti*. Naples: 1630.

[Grassi, Orazio]. *De tribus cometis anni MDCXIII Disputatio astronomica publice habita in Collegio Romano Societatis Iesu ab uno ex patribus eiusdem Societatis*. Rome: 1618.

Grendler, Marcella. "Book Collecting in Counter-Reformation Italy: The Library of Gian Vincenzo Pinelli (1535–1601)." *Journal of Library History* 16 (1981): 143–51.

Grendler, Paul F. "Books for Sarpi: The Smuggling of Prohibited Books into Venice during the Interdict of 1606–1607." In *Essays Presented to Myron P. Gilmore*, edited by Sergio Bertelli and Gloria Ramakus, vol. 1, 105–14. Florence: La Nuova Italia, 1978.

———. *The Roman Inquisition and the Venetian Press, 1540–1605*. Princeton, NJ: Princeton University Press, 1977.

———. *The Universities of the Italian Renaissance.* Baltimore: Johns Hopkins University Press, 2004.

Grevembroch, Giovanni. *Gli abiti de veneziani di quasi ogni età con diligenza raccolti e dipinti nel secolo XVIII.* Introduction by Giovanni Mariacher. Venice: Filippi, 1981. 4 vols.

Grillo, Angelo. *Lettere del molto R. P. abbate D. Angelo Grillo.* Venice: 1608.

Groebner, Valentin. *Identification, Deception, and Surveillance in Early Modern Europe.* Cambridge, MA: MIT Press, 2007.

Gualdo, Paolo. *Vita Ioannis Vincentii Pinelli, patricii genuensis.* Augsburg: 1607.

Guiducci, Mario. *Discorso delle comete.* Florence: 1619.

———. *Lettera al M. R. P. Tarquinio Galluzzi della Compagnia di Giesù.* Florence: 1620.

Hannaway, Owen. "Laboratory Design and the Aim of Science: Andreas Libavius versus Tycho Brahe." *Isis* 77 (1986): 585–610.

Harris, Steven J. "Confession-Building, Long-Distance Networks, and the Organization of Jesuit Science." *Early Science and Medicine* 1 (1996): 287–318.

Hart, Clive. *The Prehistory of Flight.* Berkeley: University of California Press, 1985.

Haskell, Francis. *Patrons and Painters: A Study in the Relations between Italian Art and Society in the Age of the Baroque,* 2nd ed. New Haven: Yale University Press, 1980.

Heilbron, John. *Galileo.* Oxford: Oxford University Press, 2010.

Henry, John. "Animism and Empiricism: Copernican Physics and the Origins of William Gilbert's Experimental Method." *Journal of the History of Ideas* 62 (2001): 99–119.

Herbst, Klaus-Dieter. "Galilei's Astronomical Discoveries Using the Telescope and Their Evaluation Found in a Writing-Calendar from 1611." *Astronomische Nachrichten* 330 (2009): 536–39.

Howard, Deborah. *Venice & the East: The Impact of the Islamic World on Venetian Architecture 1100–1500.* New Haven: Yale University Press, 2000.

———. *Venice Disputed: Marc'Antonio Barbaro and Venetian Architecture, 1550–1600.* New Haven: Yale University Press, 2011.

Hsia, Florence C. *Sojourners in a Strange Land: Jesuits and their Scientific Missions in Late Imperial China.* Chicago: University of Chicago Press, 2009.

Johns, Adrian. *The Nature of the Book: Print and Knowledge in the Making.* Chicago: University of Chicago Press, 1998.

Johnson, James. *Venice Incognito: Masks in the Serene Republic.* Berkeley: University of California Press, 2011.

Kaoukji, Natalie. "Flying to Nowhere: Mathematical Magic and the Machine in the Library." PhDd dissertation, University of Cambridge, 2003.

Kepler, Joannes. *Dissertatio cum Nuncio Sidereo.* Prague: 1610.

———. *Dissertatio cum Nuncio Sidereo / Discussion avec le Messager Céleste.* Edited and translated by Isabelle Pantin. Paris: Les Belles Lettres, 1993.

Laird, W. R. "Archimedes among the Humanists." *Isis* 82 (1991): 629–38.

Lazzerini, Luigi. "Officina sarpiana: Scritture del Sarpi in materia di Gesuiti." *Rivista di storia della Chiesa in Italia* 58 (2004): 29–80.

Legrain, [Jean] Baptiste. *Decade commencant l'Histoire du Roy Louys XIII.* Paris: 1618.

Lloyd, Christopher. *A Catalogue of the Earlier Italian Paintings in the Ashmolean Museum.* Oxford: Clarendon Press, 1977.

Locher, Johann. *Disquisitione Mathematicae.* Ingolstadt: 1614.

Lorenzini, Antonio. *Discorso dell'Ecc. sig. Antonio Lorenzini da Montepulciano intorno alla nuova stella*. Padua: 1605.

Love, Harold. *Scribal Publication in Seventeenth-Century England*. Oxford: Oxford University Press, 1993.

Maclean, Ian. "Mediations of Zabarella in Northern Germany, 1586–1623." In *Learning and the Market Place: Essays in the History of the Early Modern Book* (Leiden: Brill, 2009), 39–58.

———. *Scholarship, Commerce, Religion: The Learned Book in the Age of Confessions, 1560–1630*. Cambridge, MA: Harvard University Press, 2012.

Magini, Giovanni Antonio. *Geografia*. 2 vols. Venice: 1596–97.

———. *Tabulae secundorum mobilium coelestium*. Venice: 1585.

Malapertius, Carolus. *Austriaca sidera heliocyclica astronomicis*. Douai, France: 1633.

Malespina, Celio. *Ducento Novelle del Signor Celio Malaspina*. Venice: 1609.

Mankowski, Tadeuz. "Some Documents from Polish Sources Relating to Carpet Making in the Time of Shah Abbas." In *A Survey of Persian Art from Prehistoric Times to the Present*, edited by Arthur Upham Pope, 2431–36. London: Oxford University Press, 1939.

Marescotti, Agesilao. *Aviso sicuro contro il mal fondato auiso del signor Antonio Quirino senator Veneto*. Bologna: 1607.

Mario Infelise, "Ricerche sulla fortuna editoriale di Paolo Sarpi (1619–1799)." In *Ripensando Paolo Sarpi*, edited by Corrado Pin, 519–46. Venice: Ateneo Veneto, 2006.

Marr, Alexander. *Between Raphael and Galileo: Mutio Oddi and the Mathematical Culture of Late Renaissance Italy*. Chicago: University of Chicago Press, 2011.

Mason, Stefania, ed. "L'Inventario di Gerolamo Bassano e l'Eredità della Bottega." Special issue, *Notiziario dell'Associazione 'Amici dei Musei e dei Monumenti di Bassano del Grappa'* (2009).

Maurizio Sangalli, "Apologie dei Padri Gesuiti contro Cesare Cremonini." In *Università, accademie, gesuiti: Cultura e religione a padova tra cinque e seicento*, 77–175. Padua: Edizioni Lint, 2001. Originally published in *Atti e memorie dell'accademia galileiana di scienze lettere ed arti, parte III: Memorie della classe di scienze morali, lettere ed arti* 110 (1997–98): 241–355.

Mazza, Cristiana. "La committenza artistica del futuro doge Nicolò Sagredo e l'inventario di Agostino Lama." *Arte veneta* 51 (1997): 88–103.

———. "Frammenti inediti della scomparsa pinacoteca Sagredo." *Arte in Friuli. Arte a Trieste* 15 (1995): 133–51.

———. *I Sagredo: Committenti e collezionisti d'arte nella Venezia del Sei e Settecento*. Venice: Istituto veneto di scienze, lettere ed arti, 2004.

Megale Valenti, Anna Maria, eds. *Le Carte di Antonio Cocchi: Inventario*. Florence: Giunta Regionale Toscana, 1990.

Meietti, Roberto, and Tommaso Baglioni. *Catalogus eorum librorum omnium, qui in ultramontanis regionibus impressi apud Roberti Meietti ad insignia duorum Gallorum*. Venice: 1602.

Menini, Ottavio. *Carmina: Ad res potissimum Gallicas, Venetas, & Romanas pertinentia, varijs temporibus scripta. Eiusdem Panegyricus serenissimo principi Donato, & excelso Senatui Veneto dictus. Eiusdem De Pace, oratio. Eiusdem De Nece Henrici 4. & de Inauguratione Ludouici 13. Ad Proceres Gallos, oratio*. Venice: 1613.

Mercurii gallobelgici: Siue Rerum in Gallia & Belgio potissimum: Hispania quoque, Italia, Anglia, Germania, Polonia, vicinisque locis ab anno 1588, vsque ad Martium anni præsentis 1594, gestarum, nuncii tomus primus. Cologne: 1598.

Miller, Peter. "Friendship and Conversation in Seventeenth-Century Venice." *Journal of Modern History* 73 (2001): 1–31.

———. *Peiresc's Europe: Learning and Virtue in the Seventeenth Century*. New Haven: Yale University Press, 2000.

Mills, John. "The Coming of the Carpet in the West." In *The Eastern Carpet in the Western World from the 15th to the 17th Century*. Selected and arranged by Donald King and David Sylvester. London: Arts Council, 1983.

Moseley, Adam. *Bearing the Heavens: Tycho Brahe and the Astronomical Community of the Late Sixteenth Century*. Cambridge: Cambridge University Press, 2007.

Motsch, Andreas. "Relations of Travel, Itinerary of a Practice." *Renaissance and Reformation* 34 (2011): 207–36.

Muir, Edward. *The Culture Wars of the Late Renaissance: Skeptics, Libertines, and Opera*. Cambridge, MA: Harvard University Press, 2007.

———. *Leopold Von Ranke Manuscript Collection of Syracuse University: The Complete Catalogue*. Syracuse, NY: Syracuse University Press, 1983.

———. *Mad Blood Stirring: Vendetta and Factions in Friuli during the Renaissance*. Baltimore: Johns Hopkins Press, 1998.

Nancy, Jean-Luc. *Ego sum*. Paris: Aubier-Flammarion, 1979.

Nardi, Antonio. Review of *Gli Scritti 'De Motu Antiquiora' di Galileo Galilei: Il Ms Gal 71. Un'analisi storico-critica* by Michele Camerota. *Nuncius* 10 (1995): 808–12.

Needham, Paul. *Galileo Makes a Book*. Vol. 2 of *Galileo's O*, ed. Horst Bredekamp. Berlin: Akademie Verlag, 2011.

Norlind, Wilhelm. "Tycho-Brahe et ses rapports avec l'Italie." *Scientia: Rivista di scienzia* 49 (1955): 47–61.

Nuovo, Angela. *The Book Trade in the Italian Renaissance*. Leiden: Brill, 2013.

Nuvoloni, Laura. "Commissioni Dogali: Venetian Bookbindings in the British Library." In *For the Love of the Binding: Studies in Bookbinding History Presented to Mirjam Foot*, edited by David Pearson, 81–109. London: British Library and Oak Knoll Press, 2000.

O'Connell, Monique. *Men of Empire: Power and Negotiation in Venice's Maritime State*. Baltimore: Johns Hopkins University Press, 2009.

O'Malley, John. *The First Jesuits*. Cambridge, MA: Harvard University Press, 1993.

Pagano, Sergio. *I documenti vaticani del processo di Galileo Galilei*. 2nd edn. Vatican City: Archivio Segreto Vaticano, 2009.

Pavone, Sabina. *Le Astuzie dei Gesuiti: Le false Istruzioni Segrete della Compagnia di Gesù a la Polemica Antigesuita nei secoli XVII e XVIII*. Rome: Salerno, 2000.

Peterson, Mark A. *Galileo's Muse: Renaissance Mathematics and the Arts*. Cambridge, MA: Harvard University Press, 2011.

Petrucci, Armando. *Public Lettering: Script, Power and Culture*. Chicago: University of Chicago Press, 1993.

Pettigree, Andrew. *The Book in the Renaissance*. New Haven: Yale University Press, 2010.

Picciotto, Joanna. *Labors of Innocence in Early Modern England*. Cambridge, MA: Harvard University Press, 2010.

Piemontese, Angelo Michele. "I due Ambasciatori di Persia ricevuti da papa Paolo V al Quirinale." *Miscellanea Bibliothecae Apostolicae Vaticanae* 12 (2005): 357–425.

Pin, Corrado. "Le Scritture pubbliche trovate alla morte di fra Paolo Sarpi nel convento dei Servi." *Memorie della Accademia delle Scienze di Torino* 2 (1978): 311–69.

————. "Tra religione e politica: Un codice di memorie di Paolo Sarpi." In *Studi politici in onore di Luigi Firpo*. 2 vols. Edited by S. Rota Ghibaudi and F. Barcia. Milan: FrancoAngeli, 1990.

Placcius, Vicentius. *Theatrum anonymorum et pseudonymorum ex symbolis & collatione vitorum per Europam doctissimorum ac celeberrimorum.* Hamburg: 1708.

Poppi, Antonino. *Cremonini, Galilei, e gli inquisitori del Santo a Padova.* Padua: Centro Studi Antoniani, 1993.

[Possevino, Antonio]. *Risposta del sig. Paolo Anafesto all'auuiso del sig. Antonio Quirino.* Camerino and Bologna: 1607.

[————]. *Risposta di Teodoro Eugenio di Famagosta, all'auiso mandato fuori dal signore Antonio Quirino senatore Veneto.* Bologna: 1606.

[Possevino, Antonio?]. *Relatione della segnalata et come miracolosa conquista del paterno imperio.* [. . .] *Raccolta da sincerissimi avvisi per Barezzo Barezzi.* Venice: 1605 and Florence: 1606.

Preto, Paolo. *I servizi segreti di Venezia: Spionaggio e controspianaggio ai tempi della Serenissima.* Milan: Il Saggiatore, 1994.

Prieto, Andrès I. *Missionary Scientists: Jesuit Science in Spanish South America, 1570–1810.* Nashville: Vanderbilt University Press, 2011.

Pumfrey, Stephen. *Latitude and the Magnetic Earth: The True Story of Queen Elizabeth's Most Distinguished Man of Science.* Cambridge: Icon Books, 2002.

Radziwiłł, Mikołaj Krzysztof. *Hierosolymitana Peregrinatio.* Braunschweig, Germany: 1601.

Raines, Dorit. *L'Invention du mythe aristocratique: L'image de soi du patriciat vénetien au temps de la Sérénissime.* 2 vols. Venice: Istituto Veneto di scienze, lettere ed arti, 2006.

Rearick, William R. "More Veronese Drawings from the Sagredo Collection." *Master Drawings* 33 (1995): 132–43.

Rebellato, Elisa. *La fabbrica dei divieti.* Milan: Bonnard, 2008.

Redondi, Pietro. "I fondamenti metafisici della fisica di Galileo." *Nuncius: Annali di storia della scienza* 12 (1997): 267–90.

Reeves, Eileen. "Variable Stars: A Decade of Historiography on the *Sidereus Nuncius.*" *Galilaeana* 8 (2011): 37–72.

————. *Galileo's Glassworks.* Cambridge, MA: Harvard University Press, 2008.

Reeves, Eileen, and Albert Van Helden. *Galileo and Scheiner on Sunspots, 1611–1613.* Chicago: University of Chicago Press, 2010.

Rhodes, Dennis E. *Giovanni Battista Ciotti (1562–1627?): Publisher Extraordinary at Venice.* Venice: Marcanium, 2013.

————. *Silent Printers: Anonymous Printing at Venice in the Sixteenth Century.* London: British Library, 1995.

————. "Spanish Books on Sale in the Venetian Bookshop of G. B. Ciotti, 1602." *The Library: The Transactions of the Bibliographical Society* 12 (2011): 50–55.

Riccobono, Antonio. *De Gymnasio Patavino . . . Commentariorum Libri Sex.* Padua: 1598. Facsimile reprint Bologna: Arnaldo Forni, 1980.

Richardson, Brian. *Manuscript Culture in Renaissance Italy.* Cambridge: Cambridge University Press, 2009.

Richardson, Brian, ed. "The Uses of Manuscripts in Early Modern Italy." Special issue, *Italian Studies* 66 (2011).

Ridolfi, Carlo. *Le meraviglie dell'Arte ovvero Le vite degli illustri pittori veneti e dello Stato.* Edited by Detlev von Hadeln. Rome: Società multigrafica editrice SOMU, 1965.

Rose, Paul. *The Italian Renaissance of Mathematics: Studies on Humanists and Mathematicians from Petrarch to Galileo.* Geneva: Droz, 1975.

———. "A Venetian Patron and Mathematician of the Sixteenth Century: Francesco Barozzi (1537–1604)." *Studi Veneziani* 1 (1977): 119–80.

Rose, Stephen. "Trumpeters and Diplomacy on the Eve of the Thirty Years' War: The 'Album Amicorum' of Jonas Kröschel." *Early Music* 40 (2012): 379–92.

Rosen, Edward. "The Title of Galileo's *Sidereus Nuncius.*" *Isis* 41 (1950): 287–89.

Rota, Giorgio. "Safavid Envoys in Venice." In *Diplomatisches Zeremoniell in Europa und im Mittleren Osten in der Frühen Neuzeit,* edited by Ralph Kauz, Giorgio Rota, and Jan Paul Niederkorn, 213–45. Vienna: Verlag der Österreichischen Akademie der Wissenschaften, 2009.

——— "Safavid Persia and Its Diplomatic Relations with Venice." In *Iran and the World in the Safavid Age,* edited by Willem Floor and Edmund Herzig, 149–60. London, I. B. Tauris, 2012.

———. *Under Two Lions: On the Knowledge of Persia in the Republic of Venice (ca. 1450–1797).* Vienna: Verlag der Österreichischen Akademie der Wissenschaften, 2009.

Rothman, Nathalie. *Brokering Empire: Trans-Imperial Subjects between Venice and Istanbul.* Ithaca, NY: Cornell University Press, 2011.

[Roussel, M.] *Antimariana ou Refutation des Propositions de Mariana: Pour monstrer que la vie des Princes souverains doit ester inviolable aux subjects, & à la Republique.* Rouen: 1610.

Rubiés, Joan-Pau. "A Dysfunctional Empire? The European Context to Don Garcìa de Silva y Figueroa's Embassy to Shah Abbas." In *Estudos sobre Don García de Silva y Figueroa e os "Comentarios" da embaixada à Persia (1614–1624),* edited by Rui Loureiro and Vasco Resende, 85–133. Lisbon: Centro de História de Além-Mar, 2011.

Rule, William Harris. *History of the Inquisition.* 2 vols. New York: Scribner, 1874.

Russo, F. "Note sur la traduction du titre de l'ouvrage de Galilée, *Sidereus Nuncius.*" *Revue d'histoire des sciences* 20 (1967): 67–69.

Russo, Lucio. *The Forgotten Revolution: How Science Was Born in 300 B.C. and Why It Had to be Reborn.* Berlin: Springer, 2004. Original edition 1996.

Sak, L. N. and O. K. Shkolyarenko, eds. *Західноєвропейський живопис* (West-European Painting, 14th–18th Centuries). Kiev: Мистецтвоб, 1981.

Salmina-Haskell, Larissa. Review of *Kartini Italianskih Masterov XIV–XVIII Vekov is Museev SSSR* (*Картены Итальянских мастеров XIV–XVIII веков из Музеев СССР* [Moscow: Советский художник, 1986]) by Viktoria Markova. *Burlington Magazine* 133 (1991): 630.

Saltini, Guglielmo Enrico. "Di Celio Malespini ultimo novelliere italiano in prosa del secolo XVI." *Archivio Storico Italiano,* 5th ser., 13 (1894): 35–80.

Sansovino, Francesco and Giustiniano Martinioni. *Venetia, città nobilissima et singolare,* edited by Lino Moretti. Venice: Filippi, 1968. Facsimile reprint of the 1663 edition.

Sardella, Pierre. "Nouvelles et speculations à Venise au debut du XVIe siècle." *Cahier des Annales* 1 (1948): 1–85.

Sarpi, Paolo. *Lettere ai Protestanti.* Edited by Manlio Duilio Busnelli, 2 vols. Bari: Laterza, 1931.

———. *Scelte lettere inedite di frà Paolo Sarpi.* Capolago: Elvetica, 1833.

Sarsio, Lothario [i.e., Orazio Grassi]. *Libra astronomica ac philosophica, qua Galilaei Galilaei opiniones de comits a Mario Guiduccio in Florentina Academia expositae, atque in lucem nuper editae, examinantur a Lothario Sarsio Sigensano.* Perugia: 1619.

Savio, Pietro. "Per l'epistolario di Paolo Sarpi." *Aevum* 10 (1936): 3–104.

———. "Per l'epistolario di Paolo Sarpi (cont.)." *Aevum* 11 (1937): 13–74.

————. "Per l'epistolario di Paolo Sarpi (cont.)." *Aevum* 11 (1937): 275–322.

————. "Per l'epistolario di Paolo Sarpi (cont.)." *Aevum* 13 (1939): 558–622.

————. "Per l'epistolario di Paolo Sarpi (cont.)." *Aevum* 14 (1940): 3–84.

————. "Per l'epistolario di Paolo Sarpi (cont.)." *Aevum* 16 (1942): 3–43.

Sayce, R. A. "Compositional Practices and the Localization of Printed Books, 1530–1800." *The Library* 21 (1966): 1–45.

Schaffer, Simon. "Newton on the Beach: The Information Order of the *Principia Mathematica*." *History of Science* 47 (2009): 243–76.

Schaffer, Simon and Steven Shapin. *Leviathan and the Air-Pump: Hobbes, Boyle, and the Experimental Life*. 2nd edn. Princeton, NJ: Princeton University Press, 2011.

Schaffer, Simon, Lissa Roberts, Kapil Raj, and James Delbourgo, eds. *The Brokered World: Go-Betweens and Global Intelligence, 1770–1820*. Sagamore Beach, MA: Science History Publications, 2009.

Scheiner, Christoph. *Rosa Ursina, sive sol ex admirando facularum & macularum suarum phoenomeno varius*. Bracciano: 1626–30.

Sclosa, Meri. "Vedute possibili: Finestre paesaggistiche nella ritrattistica di Leandro Bassano e Domenico Tintoretto." *Paragone* 94 (2010): 35–52.

Scot, Alexander. *Universa Grammatica Graeca*. Lyon: 1594.

Serenissimo et Pietoso aviso della Chiesa apostolica Romana al santissimo signor Paulo Quinto Pontifice Massimo è moderno Vescouo di Roma. Sopra la differenza tra la sua santità & la sereniβima signoria di Venetia. Alquale si dimostra con chiarissime leggi è ragioni diuine è di politia Che si l'apostolico Vuol menar la guerra Non sara piu apostolico sopra la terra. [n.p.]: 1607.

Serrai, Alfredo. *Cataloghi (Tipografici. Editorali. Di librai. Bibliotecari)*. Vol. 4 of *Storia della Bibliografia*. Rome: Bulzoni, 1993.

Sestini, Domenico. *Descrizione d'alcune medaglie greche del museo del Signore Barone Stanislao di Chaudoir*. Florence: Guglielmo Piatti, 1831.

Shapin, Steven. "'The Mind Is Its Own Place': Science and Solitude in Seventeenth-Century England." *Science in Context* 4 (1991): 191–218.

Shapin, Steven. *A Social History of Truth: Civility and Science in Seventeenth-Century England*. Chicago: University of Chicago Press, 1994.

Sheehan, Jonathan. "Introduction: Thinking about Idols in Early Modern Europe." *Journal of the History of Ideas* 67 (2006): 561–70.

Shoduar, Stanislav. *Aperçu sur les monnaies russes et sur les monnaies étrangères qui ont eu cours en Russie. Depuis les temps les plus reculés jusqu'à nos jours*. 3 vols. St. Petersburg: F. Bellizard, 1836–37.

Smith, Logan Pearsall, ed. *The Life and Letters of Sir Henry Wotton*. 2 vols. Oxford: Clarendon Press, 1907.

Smolka, Josef. "The Scientific Revolution in Bohemia." In *The Scientific Revolution in National Context*, edited by Roy Porter and Mikulas Teich, 210–39. Cambridge: Cambridge University Press, 1992.

Snyder, Jon. *Dissimulation and the Culture of Secrecy in Early Modern Europe*. Berkeley: University of California Press, 2012.

Spampanato, Vincenzo. "Nuovi documenti intorno a negozi e processi dell'Inquisizione, 1603–1624 (cont.)." *Giornale critico della filosofia italiana* 5 (1924): 346–401.

Spruit, Leen. "Cremonini nelle carte del Sant'Uffizio romano." In *Cesare Cremonini: Aspetti del pensiero e scritti*, vol. 1, edited by Ezio Riondato and Antonino Poppi, 193–205. Padua: Accademia Galileiana di Scienze, Lettere e Arti, 2000.

Squitinio della liberta veneta nel quale si adducono anche le raggioni dell'Impero Romano sopra la città & signoria di Venetia. Mirandola, Italy: 1612.

Steensgaard, Niels. *Carracks, Caravans and Companies: The Structural Crisis in the European-Asian Trade in the Early 17th Century*. London: Studentlitteratur, 1973.

Stella, Aldo. "Galileo, il circolo culturale di Gian Vincenzo Pinelli e la 'Patavina Libertas.'" In *Galileo e la Cultura Padovana*, edited by Giovanni Santinello, 307–25. Padua: CEDAM, 1992.

Sturm, Johann. *Linguae latinae resolvendae ratio*. Strassburg: 1581.

Subrahmanyam, Sanjay. *Three Ways to be Alien: Travails and Encounters in the Early Modern World*. Waltham, MA: Brandeis University Press, 2011.

Szépe, Helena. "Civic and Artistic Identity in Illuminated Venetian Documents," *Bulletin du Musée hongrois des beaux-arts* 95 (2002): 59–78.

———. "Distinguished among Equals: Repetition and Innovation in Venetian *Commissioni*," In *Manuscripts in Transition: Recycling Manuscripts, Texts and Images*, edited by Brigitte Dekeyzer and Jan Van der Stock, 441–47. Leuven: Peeters, 2005.

Tarde, Jean. *Borbonia Sidera*. Paris: 1621. Translated by the author as *Les astres de Borbon*. Paris: 1622.

Targosz, Karolina. "'Le Dragon Volant' de Tito Livio Burattini." *Annali dell'Istituto e Museo di storia della scienza di Firenze* 2 (1977): 67–85.

Taton, René. "Nouveau document sur le 'Dragon Volant' de Burattini." *Annali dell'Istituto e Museo di storia della scienza di Firenze* 7 (1982): 161–68.

Thiersch, Hermann. *Pharos, Antike, Islam und Occident: Ein Beitrag zur Architekturgeschichte*. Leipzig and Berlin: B. G. Teubner, 1909.

Thompson, Jon. "Early Safavid Carpets and Textiles." In *Hunt for Paradise: Court Arts of Safavid Iran, 1501–1576*, edited by Jon Thompson and Sheila R. Canby, 270–317. London: Thames & Hudson, 2003.

Tomasini, Jacopi Philippi. *Elogia Virorum Literis*. Padua: 1644.

Ugaglia, Monica. "The Science of Magnetism before Gilbert: Leonardo Garzoni's Treatise on the Loadstone." *Annals of Science* 63 (2006): 59–84.

Valleriani, Matteo. *Galileo Engineer*. Dordrecht: Springer, 2010.

Vecellio, Cesare. *Habiti antichi, et moderni di tutto il mondo*. Venice: [1598].

Wallace, William. "The Dating and Significance of Galileo's Pisan Manuscripts." In *Nature, Experiment, and the Sciences: Essays on Galileo and the History of Science in Honour of Stillman Drake*, edited by Trevor H. Levere and William R. Shea, 3–50. Dordrecht: Kluwer, 1990.

Wedderburn, John. *Quatuor problematum quae Martinus Horky contra Nuntium Sidereum de quatuor planetis nouis disputanda proposuit. Confutatio*. Padua: 1610.

Wilding, Nick. "Galilean Angels." In *Conversations with Angels: Essays towards a History of Spiritual Communication, 1100–1700*, edited by Joad Raymond, 67–89. New York: Palgrave Macmillan, 2011.

———. "Galileo's Idol: Gianfrancesco Sagredo Unveiled." *Galilaeana: Journal of Galilean Studies* 3 (2006): 229–45.

———. "Science and the Counter-Reformation." In *The Ashgate Companion to the Counter-Reformation*, edited by Mary Laven, Alexandra Bamji, and Geert Janssen, 319–36. Aldershot, UK: Ashgate, 2013.

———. "The Strangest Piece of News." Review of *Galileo: Watcher of the Skies*, by David Wootton, and *Galileo*, by John Heilbron. *London Review of Books*, 2 June 2011, 31–32.

Wilkins, John. *Mathematicall magick*. London: 1648.

Wootton, David. *Galileo: Watcher of the Skies*. New Haven: Yale University Press, 2010.

———. *Paolo Sarpi: Between Renaissance and Enlightenment*. Cambridge: Cambridge University Press, 2002.

Wotton, Henry. *Reliquiae Wottonianae*. London: 1672. First edition 1651.

Woudhysen, H. R. *Sir Philip Sidney and the Circulation of Manuscripts, 1558–1640*. Oxford: Oxford University Press, 1996.

Yale, Elizabeth. "Marginalia, Commonplaces, and Correspondence: Scribal Exchange in Early Modern Science." *Studies in History and Philosophy of Biological and Biomedical Sciences* 42 (2011): 193–202.

Zambella, Paola. "Uno, due, tre mille Menocchio?" Review of Carlo Ginzburg *Il Formaggio e i vermi: Archivio storico italiano* 137 (1979): 51–90.

Zemon Davis, Natalie. *The Return of Martin Guerre*. Cambridge, MA: Harvard University Press, 1984.

———. *Trickster Travels: A Sixteenth-Century Muslim between Worlds*. New York: Hill & Wang, 2006.

Zilsel, Edgar. "The Origins of William Gilbert's Scientific Method." *Journal of the History of Ideas* 2 (1941): 1–32.

Index

Page numbers in italics refer to figures and plates.